D0961851

COME ON, MAN!

COME ON, MAN!

The Truth About Joe Biden's Terrible, Horrible, No-Good, Very Bad Presidency

JOE CONCHA

BROADSIDE BOOKS

HarperCollins books may be purchased for educational, business, or sales promotional use. For information, please email the Special Markets Department at SPsales@harpercollins.com.

FIRST EDITION

Library of Congress Cataloging-in-Publication Data
Names: Concha, Joe, author.
Title: Come on, man!: the truth about Joe Biden's terrible, horrible, no-good, very bad presidency / Joe Concha.
Other titles: Truth about Joe Biden's terrible, horrible, no-good, very bad presidency
Identifiers: LCCN 2022022138 (print) | LCCN 2022022139 (ebook) | ISBN 9780063276123 (hardcover) | ISBN 9780063276130 (ebook)
Subjects: LCSH: Biden, Joseph R., Jr. | United States—Politics and government—2021–| Presidents—United States—History—21st century. | Personality and politics—United States.
Classification: LCC E916 .C66 2022 (print) | LCC E916 (ebook) | DDC 973.934—dc23/eng/20220525
LC record available at https://lccn.loc.gov/2022022138
LC ebook record available at https://lccn.loc.gov/2022022139

22 23 24 25 26 LSC 10 9 8 7 6 5 4 3 2 1

For everyone who had the guts to pursue their passion,
despite long odds, and succeeded.

You're betraying your whole life if you don't say what you think—and you don't say it honestly and bluntly.

—*Charles Krauthammer*

CONTENTS

INTRODUCTION

"Don't underestimate Joe's ability to fuck things up."

—Barack Obama (reportedly)

Underpromise. Overdeliver. It's the number one rule in politics and sales and buying a wedding ring.

Joseph R. Biden Jr. has never embraced this simple rule.

"If you hear nothing else I say tonight, hear this. ANYONE that is responsible for that many deaths should not remain as president of the United States of America," candidate Biden declared during the first presidential debate against President Trump in September of 2020 regarding the novel coronavirus.

"I will take care of this," he declared. "I will END this. I'm going to SHUT DOWN the virus, not the country."

What kind of person promises to shut down a virus? Did Joe Biden, who had underperformed as a student, senator, two-time presidential candidate, and vice president know something that leaders of every country on the planet *didn't know*? Was there some kind of genius plan in store that COVID would never see coming?

To quote the great philosopher Mike Tyson after he was asked if he was concerned about his opponent's fight plan going into a bout:

"*Everyone has a plan until they get punched in the mouth.*"

And in Biden's first year as president, he got punched by reality and changing conditions on the ground. A lot.

By the way, if Biden had any standards, *he* would have resigned already based on his own criteria. Because more people died of

COVID in 2021 under Biden (*with* vaccines and therapeutics available, mind you) than they did the previous year under Trump (*without* vaccines and therapeutics available, mind you).

Anyone who is responsible for that many deaths should not remain as president of the United States of America, after all. They're your rules, old sport.

But as this president and this administration have done so often, in Afghanistan, with inflation, they *swore* that Omicron variant came without warning in playing the victim of circumstances card (a common theme you'll see throughout this book).

"I don't think anybody anticipated the Omicron variant," Biden said on Dec. 21, 2021. "It all started all of a sudden."

Vice President Harris echoed that inspiring sentiment.

"We didn't see Delta coming. I think most scientists did not—upon whose advice and direction we have relied—didn't see Delta coming," Harris said on December 17, 2021. "We didn't see Omicron coming. And that's the nature of what this, this awful virus has been, which as it turns out, has mutations and variants."

It "turns out" a virus "has mutations and variants"??? Are these really the top two people in charge? Biden's *own testing experts* warned a surge was coming. I'm no doctor, but I can read. And anyone who has been alive and is literate knows that every virus has mutations and variants.

Team Biden would eventually announce in January that at-home tests were on the way. And when they finally did arrive, weeks later, Omicron had spread so rapidly that it had nowhere else to go and petered out. By the end of February, daily cases in the U.S. were below 50,000, a relatively low number when considering the country was averaging more than 500,000 one month prior. The billions spent on the testing were essentially wasted since the tests were largely no longer needed.

The lesson here is that no one person, especially someone as inept as Joe Biden has proven to be throughout his entire existence as a lawmaker and candidate, should ever promise to control a virus. *To stop a virus.* And when this president says he's going to conquer cancer, it's such utter bullshit to the point of insulting by playing with people's emotions like that for a cheap applause line in somehow attempting to appear magnanimous.

By every metric, America has arguably the worst president of the television era residing in the Oval Office (at least when he isn't in Delaware). This isn't some bloviating opinion, by the way. It's based on basic facts:

Inflation at a forty-year high. Gas prices are at their highest point ever. Murders at a twenty-five-year high. Illegal border crossings at an all-time high. Parents are furious over an increasingly extremist "education" agenda for our kids. The world is increasingly unstable with our enemies more brazen, more emboldened, more dangerous.

We'll go into more detail on these metrics later, but it's truly hard to see where this president is succeeding or even shooting par.

Democrats and the media insisted up and down for five years before Joe Biden became President Biden that Trump would lie like a person breathes. They told us his kids were corrupt. They said he didn't care about immigrants despite being married to one. They said he spit on our soldiers' graves. They claimed he was a cognitive mess, that his rhetoric would cause World War III. He would start a worldwide recession (here's looking at you, Paul Krugman), and that he would only talk to "state-run media." And he would embarrass us abroad while keeping a light schedule.

All of this sentiment and these portrayals were correctly applied to the U.S. president, that's for certain. Except they don't apply to Trump, but to his successor.

So the questions many have about the current president (and none of them are comforting) are:

Is Joe Biden allowing his presidency to be run by other people, his chief of staff Ron Klain, his Domestic Policy Advisor Susan Rice, his former boss Barack Obama, his wife Jill Biden, because he's mentally checked out? Or is he your typical DC swamp creature who "evolves" (the media's term for Democrats who flip-flop) on issues as polling on said issue shifts?

Is Joe Biden *really* this liberal by insisting on expanding government to unprecedented levels, spending our way out of inflation, opening our borders, and teaching kids more about race and sex than reading, writing, science, and math?

Whose ship is this anyway? And are many in the media covering for Joe Biden by making excuses for his myriad of failures because they think their viewers and readers are stupid/uninformed? Or because they're fooling themselves thanks to the conformity of thought that exists in the political media hubs of deep (deep) blue New York City and Washington?

And finally, is Team Biden just plain unlucky or is everything Americans are seeing and feeling actually the administration's fault?

According to the president, everything that has gone wrong in this country is everyone else's fault. Inflation is on Putin. Gas prices? Also Putin, along with evil U.S. oil companies. The border? Trump left it wide open, according to his successor. It's enough to make your hair hurt.

"Here's my promise to you: If I'm elected president, I will always choose to unite rather than divide. I'll take responsibility instead of blaming others. I'll never forget that the job isn't about me—it's about you," Biden promised right before the election in a tweet that has aged as well as mayonnaise left out in the sun.

Here's what we do know: We have Joe Biden because many in

media moved heaven and earth to make him the forty-sixth president. They felt it was their moral duty to save us from Donald Trump. The guy who replaced Trump was seen as irrelevant to the overall mission. And with that myopic focus on ousting the forty-fifth president, they never bothered to scrutinize or challenge the forty-sixth.

And when it comes to Joe Biden, there was and is *plenty* to scrutinize . . .

COME ON, MAN!

1

THE MAN WHO BROKE THE BORDER

The Biden administration's policy on border defense has been an absolute fiasco. There are hundreds of questions one could ask about Joe Biden's knuckleheaded immigration policy, but leave it to our intrepid activist media superstars to start with the most stupid one.

Is Joe Biden just *too* tough on immigrants?

The question to candidate Biden came from open borders activist Jorge Ramos, who plays a serious journalist on TV.

It was September 12, 2019. The Democratic primary debate was being held at Texas Southern University, in Houston.

Ramos began with this:

> Let me start with an issue that is causing a lot of division in this country: immigration. Vice President Biden, as a presidential candidate, in 2008, you supported the border wall, saying, "Unlike most Democrats, I voted for 700 miles of fence." This is what you said.
>
> Then you served as vice president in an administration that deported 3 million people, the most ever in U.S. history. Did you do anything to prevent those deportations? I mean, you've been asked this question before and refused to answer, so let me try once again. Are you prepared to say tonight that you and President Obama made a mistake about deportations? Why should Latinos trust you?

You read all of that correctly. ABC News actually chose this person to be a debate moderator. This "journalist" scolded Biden for not doing anything to "prevent" deportations. The journalist was also saying deportations of those who entered the United States illegally was a "mistake." The journalist then lumped the entire Latino community in with his radical views when talking about trust. I wonder what Comrade Ramos thought of the surge of Latinos voting for Trump in 2020?

Biden, who flip-flops more than your average jellyfish, eventually answered this way: "I would, in fact, make sure that there is, we immediately surge to the border. All those people seeking asylum they deserve to be heard, that's who we are, we're a nation that says if you want to flee and you are fleeing oppression you should come," the candidate—who once loved border barriers—declared.

"Surge to the border." We haven't seen an open invitation like this since the prom queen invited an entire high school back to Jake Ryan's house in *Sixteen Candles*.

Here's something you should note: Biden may seem like a doddering idiot, stumbling from one mistake to the next, but it's funny how those mistakes always hew closely to liberal dreams for American policy. The president may not always know what he's saying until he's said it, but you can bet he believes what he says. His gaffes spring from the fact that, in contrast to his popular image as an ordinary, even moderate guy, he's the same elitist liberal ideologue he has been for his entire career, unless being one is politically inconvenient in the moment. And if you asked an average Biden staffer how his administration was going, they'd probably look at the hundreds of thousands of border crossings and say, "Everything's going according to plan."

Fast-forward to President Biden's first formal solo press conference at the White House in March 2021. And if you think Jorge

Ramos is just about the most biased journalist ever to ask Biden questions, allow me to introduce NBC's and PBS's Yamiche Alcindor, somehow an award-winning White House correspondent who also injects into her "questions" every far-left policy position in the book.

Even Alcindor couldn't avoid the reality that Biden's policy had been disastrous for the rule of law, but she spun the question so hard Babe Ruth would have been impressed. Alcindor "confronted" Biden: "You've said over and over again that immigrants shouldn't come to this country right now; this isn't the time to come. That message is not being received. Instead, the perception of you that got you elected—as a moral, decent man—is the reason why a lot of immigrants are coming to this country and entrusting you with unaccompanied minors."

Wait. Millions are picking up everything and possibly risking their lives by migrating here because Joe Biden was elected as a moral, decent man? Millions are entering this country illegally because the presidential version of Peyton Manning is in the White House?

Or maybe, I don't know, they simply heard Biden's invitation, along with calls from most Democrats, to come to this country illegally. These poor folks were told they will be not only welcomed, but provided free health care and housing! And they may have also heard Democrats describe the border itself as immoral and racist.

Biden, unsurprisingly, took the question as a chance to preen. "Well, look, I guess I should be flattered people are coming because I'm the nice guy; that's the reason why it's happening—that I'm a decent man or however it's phrased. That—you know, that's why they're coming, because they know Biden is a good guy."

But then he did a trademark pivot to "real talk." Only, the real talk was a shrug. The president explained that illegal border crossings were par for the course. Nothing to see here. The numbers, he argued, were similar to what they were under Trump.

"It happens every single, solitary year: There is a significant increase in the number of people coming to the border in the winter months of January, February, March. That happens every year."

Not at this pace, sir. Oh, sorry, that's what a real journalist would have said. But Yamiche wasn't finished with her speeches, er, questions. She pivoted to another pet subject.

"When it comes to the filibuster, immigration is a big issue, of course, related to the filibuster, but there's also Republicans who are passing bill after bill trying to restrict voting rights. [Senate Majority Leader] Chuck Schumer's calling it an existential threat to democracy. Why not back a filibuster rule that at least gets around issues, including voting rights or immigration? [South Carolina congressman] Jim Clyburn, someone, of course, who you know very well, has backed the idea of a filibuster rule when it comes to civil rights and voting rights."

That's not even bias in broad daylight. That's *outright activism* in pressing the president on national television to move forward with abolishing the filibuster to advance an agenda *she* supports. Alcindor also accuses Republicans of restricting voting rights. She then quotes (checks notes) *Chuck Schumer* in framing the "attack on voting rights" by eeeeeeeeeeevil GOPers as "an existential threat to democracy."

This could have been Joy Reid or Seth Meyers speaking and it would have been impossible to tell the difference.

Biden then leveled this doozy: "The idea that I'm going to say, which I would never do, if an unaccompanied child ends up at the border, we're just going to let them starve to death and stay on the other side, no previous administrations did either, except Trump."

Yup. Biden accused Trump of *starving children to death at the border*, which obviously did not happen. Amazingly, neither Yamiche nor any other reporter in the room called upon Biden to challenge him on this ridiculous accusation.

Overall, during his first press conference in 2017, President Trump was interrupted sixteen times, while Biden was interrupted here four times during his first.

When all was said and done, in Biden's first year, at least more than 2.1 million people crossed into this country illegally (those are the ones we know about, hence "at least").

In a related question, have you ever been to Denver? Major city. It has an NFL, NBA, NHL, MLB, and MLS team. If you want to know what 2.1 million is in context, take Denver's total population and triple it. Congrats, Team Biden, you just added three Denvers of potential voters to the rolls at a future election in this country. And in just one year!

Noncitizens will never be allowed to vote, you say? It's already happening in places like New York City, where a law was passed in 2021 allowing *800,000 noncitizens* to vote in local elections. Overall, at least fifteen cities and towns across the U.S. allow noncitizens to vote in local elections, from cities like San Francisco to some small towns in Vermont. And the number and list will only continue to grow.

Just don't call what's happening at the border a crisis. Yup, to avoid raising any eyebrows among most of the press and general population, the White House devised a communications strategy to deny the existence of an out-of-control surge at the border. Even President Biden's homeland security secretary, Alejandro Mayorkas, engaged in this nonsense when taking questions from reporters in March 2021.

"The men and women of the Department of Homeland Security are working around the clock seven days a week to ensure that we do not have a crisis at the border—that we manage the challenge, as acute as the challenge is," Mayorkas said on March 1.

Ah, Okay. It's a *challenge. Not* a crisis. Which completely explains why the administration was forced to open the very migrant facilities that candidate Biden once called during the Trump era "horrifying scenes at the border of kids being kept in cages" and candidate Kamala Harris described as "babies in cages," which constituted "a human rights abuse being committed by the United States government."

But fear not, because then–White House press secretary Jen Psaki explained that the very same migrant facilities that housed "babies in cages" no longer exist under the current administration.

When challenged about the double standard by Peter Doocy in February 2021, Psaki became defensive and patronizing simultaneously.

"I'm sure you're not suggesting that we have children right next to each other in ways that are not COVID safe. Are you?" she asked.

"This is not kids being kept in cages," she later insisted while adding that the reopening was only temporary. "This is a facility that was opened that's going to follow the same standards as other HHS facilities. It is not a replication. Certainly not."

Except it was.

Migrants overwhelmed facilities at a time when vaccine distribution was in its early stages. One migrant facility in Donna, Texas, alone was at *729 percent pandemic capacity* at the beginning of March 2021.

"Some of the boys said that conditions were so overcrowded that they had to take turns sleeping on the floor," Neha Desai, an attorney who represents migrant youth in U.S. custody, told CBS News at the time.

In an effort to block these images from ever being seen by the American public outside of Fox News and a few other outlets, U.S. Customs and Border Protection barred the press from photographing or filming them in the first few months of Biden's term.

This prompted award-winning Getty photographer John Moore to take to Twitter to plead for transparency. "I respectfully ask US Customs and Border Protection to stop blocking media access to their border operations. I have photographed CBP under Bush, Obama and Trump but now - zero access is granted to media. These long lens images taken from the Mexican side."

This is honestly so confusing. Gone was the utopian praise for the administration that we heard in early 2021. We were told by so many in the media that the days of transparency and truth had returned when Biden took office. One would think the journalism community would be outraged over the media blackout at the border. Instead, it was mostly silence of the lambs.

"Why have we still not seen any images inside these facilities?" one reporter asked Psaki at a press briefing in March 2021.

"The DHS oversees the Border Patrol facilities, and we want to work with them to ensure we can do it, respecting the privacy and obviously the health protocols, required by COVID," Psaki responded (without laughing).

The reporter followed by asking why photos hadn't been released from the facilities yet.

"Again, we remain committed to sharing . . . data on the number of kids crossing the border, the steps we're taking, the work we're doing to open up facilities, our own bar we're setting for ourselves in improving and expediting the timeline and the treatment of these children," Psaki explained in another nonanswer before audaciously making this claim: "We remain committed to transparency."

As things got worse and the numbers increased, Psaki even had the audacity to say Trump's border wall was never going to be effective even when totally completed.

"If we just dial it back a few years to kind of what we inherited here, the former president invested billions of dollars in a border wall that was never going to work or be effective," she actually argued in April 2022. The hubris and dishonesty was just breathtaking.

That update never came, largely because most of the national press decided the border wasn't much of a story at all. Another reason was that it was an effort to protect Biden on an issue that he and his team clearly were not handling well.

The good news is that the American people don't rely solely on broadcast news networks or CNN or the *New York Times* or *Washington Post* for their news. As a result, Biden-Harris polled in the *20s* on their handling of the border throughout 2021.

In 2022, the crisis has only gotten more dire, if that's possible. In May, a plot to assassinate former President George W. Bush was thwarted by the FBI. The plot was to be carried out by smuggling a team across the southern border. Overall, 42 terror suspects have been apprehended at the border since Biden took office. One can only wonder how many have entered the country and haven't been caught.

But activist journos like Jorge Ramos are likely thrilled to see the dream of an open border coming closer to a reality.

What likely isn't good news for Ramos is the polling numbers coming out of the Hispanic community: According to the Pew Research Center, just over one-third of Hispanics (36 percent) say the Biden administration is doing a good job handling the influx of illegal migrants entering the country, while a majority (53 percent) say immigration needs to undergo "major changes."

But here's the most alarming number for Democrats: While Biden won 61 percent of the Hispanic vote in 2020, he was polling at 26 percent in a Quinnipiac poll just fifteen months after taking office. More than half of his support, gone. Whoops.

Again, it's worth underlining that on Planet Biden, everything is going according to plan. You and I, being sane people, look at an overwhelmed border and see a ton of social and economic problems coming down the pike. We see the rule of law flouted and authority looking the other way as smugglers and traffickers conceal themselves among crowds of refugees charging the border. But Joe Biden looks at this and thinks . . . the plan's going well. Biden may be confused and overwhelmed by events of the day, but when people are confused they tend to revert to their most basic instincts. Joe Biden's most basic instinct is to be a liberal ideologue.

An unspoken consequence of this border crisis is a deadly one, because something else besides people is coming over the border and is killing younger Americans in record numbers.

Fentanyl.

2

FENTANYL

The Crisis No One's Talking About

Question: What kills more Americans between the ages of 18 and 49 than anything else in the country?

a) Cancer
b) Suicide
c) Car crashes
d) COVID
e) None of the above

If you answered E, you are correct. Instead, the answer is "F," as in fentanyl, according to the recent data from the Centers for Disease Control and Prevention (CDC).

But while COVID and Ukraine and the midterm elections have invariably led our news cycles for the past two years, the disturbing opioid crisis driven by fentanyl has received relatively little coverage.

How can that be? Here you have a drug that killed more than 100,000 Americans in 2021 alone and hospitalized hundreds of thousands more. Families are being torn apart. Rich, poor, middle class, doesn't matter. And the emergence of counterfeit oxycodone pills laced with fentanyl is fueling more deaths.

Take the deaths of two high school students that occurred within a twenty-four-hour period in Portland, Oregon, for example. Griffin Hoffmann, sixteen, was a sophomore at McDaniel High School. Already the top player on the school's tennis team. Smart. Never in trouble. He was found dead on March 6, 2022, at his desk at home with his earbuds on and his laptop open, according to police.

Then there's Olivia Coleman, seventeen, a junior at McDaniel. "Loved children and often babysat her cousins. She was known for her empathetic, nurturing nature. She talked about wanting to become a child psychologist someday," according to the *Oregonian*. She was found dead on March 7, 2022, in her bedroom.

According to investigators, Griffin and Olivia likely thought they were taking pills consisting of oxycodone, but the ones they took were made of fentanyl. The pills taken were blue and marked "M30," which indicates oxycodone.

According to the CDC, fentanyl is a potentially fatal synthetic opioid that, in some cases, can be fifty times more potent than another opioid, heroin. *Fifty times.*

The trends should worry everyone: The Drug Enforcement Administration (DEA) reports, "[O]verdose deaths involving synthetic opioids (primarily illicitly manufactured fentanyl) rose 55.6 percent [last fiscal year] and appear to be the primary driver of the increase in total drug overdose deaths."

So if that's the case, the United States is looking at more than 150,000 American opioid overdose deaths in 2022—enough to fill Yankee Stadium three times over.

For whatever reason, the Biden administration and many lawmakers don't seem to be addressing the issue in any meaningful way. In 2022, there's been too much talk about spending trillions to build back better and federalizing voting rights. Because why outline a comprehensive fentanyl and drug overdose strategy to save countless lives when apparently "voting rights" and democracy are in peril and will die unless the filibuster is blown up?

President Biden has almost never spoken personally about this crisis, nor has Vice President Harris. Back in November 2021, the president issued a boilerplate, generic statement that barely generated any press coverage.

"We are strengthening prevention, promoting harm reduction, expanding treatment, and supporting people in recovery, as well as reducing the supply of harmful substances in our communities. And we won't let up," the statement reads.

But never judge any president or Congress on their words. Look at their deeds and actions instead. And in this case, Congress earmarked nearly $4 billion to address mental health and substance use disorder programs. Looks good on paper, right? Until you consider that trillions upon trillions have been spent in addressing all things COVID.

In a world where the U.S. spent nearly $7 trillion in 2021, $4 billion ain't much.

While the president said the administration is "reducing the supply of harmful substances in our communities," his speechwriters failed to mention how record amounts of fentanyl entered the country in 2021 through the seemingly wide-open U.S. southern border.

So how bad is this fentanyl crisis?

According to U.S. Customs and Border Protection (CBP), there was a *1,066 percent increase* in fentanyl seized in 2021 compared to 2020. And that's what was *seized*. Given that "fewer than five percent of vehicles are typically screened by CBP with the powerful high-energy scanners that can peer deep inside cargo loads to detect anomalies—odd patterns or suspicious densities that could be illegal drugs," according to the *Washington Post*, there is more than enough fentanyl in the United States today to kill every citizen of this country.

So, will the president visit the southern border where all of this deadly stuff is coming in? I wouldn't hold my breath. It's not really a priority on Planet Biden. Here's what he told CNN in a town hall in October 2021:

"I've been there before and I haven't—I mean, I know it well. I guess I should go down. But the whole point of it is I haven't had a whole hell of a lot of time to get down," he told Anderson Cooper, who apparently didn't feel any need to challenge Biden on his untruth.

For starters, Biden has barely been to the southern border despite being a lawmaker for fifty years. If you want to count that he drove by it, kinda, briefly in a motorcade once in 2008 on his way from a Texas airport to a campaign rally located one hour from the border in New Mexico, so be it.

In fact, here's Jen Psaki actually making the argument that somehow *this counts* as a border visit when she was asked by Fox's Peter Doocy if Biden had even been there.

> As you may have seen, there's been reporting that he did drive through the border when he was on the campaign trail in 2008. And he is certainly familiar with the fact—and it stuck with him—with that fact that in El Paso, the border goes right through the center of town. . . . He does not need a visit to the border to know what a mess was left by the last administration.

And there it is! The White House press secretary again blamed Trump for the current crisis.

Thankfully, Doocy wouldn't accept this comical answer.

"Does that count as a visit? He said, 'I've been there before'—you're saying he drove by for a few minutes—does that count?" he rightly asked.

"What do you—what is the root cause?" Psaki responded without answering the question before going into the usual condescension mode. "Where are people coming from who are coming to the border, Peter? I'm asking you a question because I think people should understand the context."

Doocy didn't take the bait in becoming the story by answering Psaki's question.

"He said, 'I guess I should go down.' So does he think that he needs a photo op?" he asked.

"Okay, I'll answer it for you," Psaki said of her own question. "People come from Central America and Mexico to go to the border. The president has been to those countries ten times to talk about border issues.

The last time he went to Central America, and it wasn't to discuss border security, was in (checks notes) 2014.

"The president does not believe a photo op is the same as solutions," Psaki added.

As for visiting as president, that's unlikely to happen because there's no political upside for doing so.

Oh, by the way, do you know where fentanyl primarily comes from?

Let's do another multiple choice:

a) China
b) China
c) China
d) China
e) All of the above

Yep, that's right. China is the primary supplier of the number one death drug in America. China, inadvertently or not, sent COVID all over the world, killing millions, including more than one million Americans. We shut down the country—schools, businesses, sports, sanity—because of it.

Now China is sending another killer our way. And thanks to Joe Biden and border czar Kamala Harris, they have the perfect point of entry to get it into the country: the porous U.S.-Mexico border.

"The role of China—and the growing role of India—is sending precursor chemicals directly to cartels in Mexico to produce the fentanyl in clandestine labs," Matt Donahue, the DEA deputy chief of operations, told NPR shortly after Biden was elected. "You push on a balloon, it pops somewhere else." That pop is occurring across all fifty U.S. states at an alarming rate.

As the deaths mount, especially among young, promising teens like the late Griffin Hoffmann and Olivia Coleman, this crisis will be impossible to ignore. The media—especially the big boys like *60 Minutes*—needs to cover this gigantic story more and shine a floodlight on the current fentanyl crisis more instead of the Washington insider crap we get to ingest on a daily basis. Because when something is the leading killer of an entire demographic, it warrants serious attention. Or at least one would think.

It would also be nice if a president, who has seen firsthand what drug abuse and addiction look like courtesy of his son Hunter, would make this crisis a top priority.

But even if Biden did focus on fixing this problem, would anyone trust he has the competency to do so? Why does the president who portrays himself as a champion of the working class not care?

The reason? Fixing the problem would require taking responsibility for making it infinitely worse by his opening of the border. To help slow the flow and arrest those dealing this stuff, it would take a complete 180 on border and crime policy. That would alienate the left. To Joe Biden, that simply can't happen. It's a prospect this president prioritizes over serving and protecting the public. That's a whole fatal bowl of wrong.

Speaking of wrong, our commander in chief couldn't be any more wrong than he was on the U.S. withdrawal from Afghanistan. Its execution, or lack thereof, would mark the turning point in his presidency.

3

FOREIGN POLICY AMATEUR HOUR

July 8, 2021

The president of the United States is announcing from the East Room of the White House a timeline for a drawdown of U.S. forces in Afghanistan. The military mission of nearly two decades would end on August 31, according to Joe Biden, regardless of conditions on the ground.

After making his remarks regarding said drawdown, Biden was asked the following question: "Mr. President, some Vietnamese veterans see echoes of their experience in this withdrawal in Afghanistan. Do you see any parallels between this withdrawal and what happened in Vietnam?"

"None whatsoever," Biden shot back without hesitation. "Zero."

He continued to ramble, probably hoping people would forget the question by the end of his word salad. He said, "What you had is you had entire brigades breaking through the gates of our embassy—six, if I'm not mistaken. The Taliban is not the South . . . the North Vietnamese army. They're not—they're not remotely comparable in terms of capability. There's going to be no circumstance where you see people being lifted off the roof of an embassy in the—of the United States from Afghanistan. It is not at all comparable."

This is Biden at his craptastic finest. It's difficult to fully illustrate just how wrong he's been on almost every foreign and defense policy. For those who ever saw the 1993 classic *A Bronx Tale*, Biden is Eddie Mush, an addicted gambler who loses every bet he makes. Even the narrator roasts Mush as "the biggest loser in the whole world."

In the movie, there's a hilarious scene at the racetrack where a mob boss named Sonny, played by the great Chazz Palminteri, is watching his appropriately named horse Kryptonite enjoy a sizable lead entering the homestretch.

"It's a lock," Sonny says while shaking hands with his associates. "Don't even worry about it. We can't lose."

But then the crew spots Mush loudly cheering on the same horse. Sonny tears up his tickets, even though Kryptonite is still ahead by five lengths, and leaves the grandstand before the race is even over.

"We can't win! Mush bet Kryptonite!" Sonny declares. "We've been mushed."

Kryptonite was indeed mushed. He would go on to lose by a nose.

In 1991, Joe Biden, playing the role of Senator Mush (D-DE), voted against the Gulf War, a relatively brief conflict that achieved its primary goal of removing Saddam Hussein and Iraq from Kuwait after the dictator invaded it for oil.

That war, by the way, was supported by then-senator Al Gore (a future Democrat vice president) and Joe Lieberman (a future Democrat vice presidential candidate). Oh, no worries: Mush/Biden would get a do-over in 2003 when he reversed his approach and voted in favor of the Iraq War. Turns out that was the wrong war to support (Donald Trump publicly opposed it at the time). There were no weapons of mass destruction, despite what we were told. Plus, Saddam had nothing to do with the terrorist attacks of September 11, 2001.

Mush/Biden, who seemingly doesn't realize that statements and votes were recorded in 2003, would swear as a 2020 presidential candidate that he never voted for the Iraq War.

"From the moment 'shock and awe' started, from that moment, I was opposed to the effort, and I was outspoken as much as anyone at all in the Congress and the administration," Biden said during a July 2020 primary debate.

The moderator, Jake Tapper, didn't fact-check Biden on this in real time. But here's what Mush/Biden told CNN on March 19, 2003, right after the war began.

> We have one single focus. And that is, we're about to send our women and men to war. The president is the commander in chief. We voted to give him the authority to wage that war. We should step back and be supportive.

And:

> There's a lot of us who voted for giving the president the authority to take down Saddam Hussein if he didn't disarm. And there are those who believe, at the end of the day, even though it wasn't handled all that well, we still have to take him down.

Senator Mush was also the brains behind a 2006 plan to divide Iraq into ethnic and sectarian enclaves (Sunni, Shiite, and Kurdish zones), which was widely rejected by Iraqi leaders. Sectarian killings were driving Iraq into chaos at the time, and Biden's plan to create borders within the country would only escalate the crisis.

But most notably, it was Vice President Mush/Biden who opposed the raid by U.S. special forces to kill Osama bin Laden at a compound where he was hiding out in the city of Abbottabad, Pakistan.

On May 2, 2011, Mush joined President Obama and his top national security officials for a top-secret meeting about intelligence that bin Laden was finally located and could be taken out.

"Joe, what do you think?" Obama asked his vice president on whether SEAL Team Six should go in or not.

"Mr. President, my suggestion is don't go," Mush replied.

Obama ignored his alleged BFF and sent in special forces anyway. The terror leader was killed without any U.S. casualties. The world would be a very different place if Obama hadn't put in the order to whack Osama, especially politically here at home. Remember, Obama-Biden's reelection slogan was "GM is alive and bin Laden is dead." Pithy. Effective. Easy to absorb. But one has to wonder if Obama would have still won in 2012 without the second part of that bumper sticker, especially if Al Qaeda pulled off another major terror attack with bin Laden appearing on video thumbing his nose at the U.S. after it.

Fast-forward to January 2021, and Mush had somehow elevated himself to commander in chief. Given this sterling track record, what could possibly go wrong?

So let's return to July of that year: Mush tells the American people how well the Afghan army has been trained. He praises how they have defended the government friendly to the U.S. from the Taliban. This will be like Tom Brady's Bucs taking on the Jaguars, Mush basically was predicting, with the Taliban playing the role of the hapless, overwhelmed Jags.

"We have trained and equipped over three hu—nearly 300,000 current serving members of the military—of the Afghan National Security Force, and many beyond that who are no longer serving," Mush declared. "Add to that, hundreds of thousands more Afghan

National Defense and Security Forces trained over the last two decades.

"We provided our Afghan partners with all the tools—let me emphasize: all the tools, training, and equipment of any modern military. We provided advanced weaponry. And we're going to continue to provide funding and equipment. And we'll ensure they have the capacity to maintain their air force," he continued.

Well, it sounded like we really had our shit together on this one. And if Biden and his handlers are accomplished at anything, it's reading polls and governing accordingly. In the summer of 2021, 7 in 10 Americans supported our getting out of Afghanistan. Those Americans likely felt that way, at least in retrospect, because the president assured them the transition would be without cost. Without thirteen U.S. service members being murdered. Without Americans being left behind. Without us looking weak and incompetent to our adversaries in China, Russia, Iran, and North Korea.

As we all know, the outcome was a disaster. The Taliban, which will never be confused with the Chinese military, seized city after city and province after province with little resistance, including the capital city of Kabul in about the time it takes to deliver a pizza. Consequently, the U.S. embassy was rapidly evacuated via helicopter, just as Mush predicted would never happen. The photo comparisons between Kabul in 2021 and Saigon in 1975 were almost indistinguishable.

"What is abnormal is the scale of American helicopters circulating around the area of the embassy," CNN International's Nick Paton Walsh reported at the time. "I have not seen anything like this in twenty years, in terms of the volume."

Team Biden got hit by both conservatives and liberals for its miscalculation that the Taliban could be reasoned with, or fought off internally, and would not immediately proceed to take back the country, including the capital city of Kabul. Everyone was asking the same questions: Why did Biden announce the departure of U.S. military personnel at the onset of the fighting season instead of completing the process in the dead of winter, when the Taliban retreat to their bases in neighboring Pakistan? Why did the U.S. leave Bagram Air Base six weeks before its final withdrawal? If Bagram was available and secure, we could have used it to evacuate Americans and Afghan allies out of the country instead of relying on the *friggin' Taliban* to provide security at Kabul International Airport. The Bagram decision would lead directly to those thirteen American service members going home in caskets.

So why did Biden make that decision? It was probably because of a behind-the-scenes deal, made out of panic, according to a former special army special forces veteran who spoke to the *New York Post*. "I think the Taliban demanded [the closure of Bagram]," Jim Hanson, told the paper in August 2021. "I think that was part of the deal that Biden made. The Taliban threatened to start fighting again and then Biden got scared."

For his part, Mush passed the buck and blamed his military advisers for the decision. "They concluded—the military—that Bagram was not much value added, that it was much wiser to focus on Kabul. And so, I followed that recommendation," he explained.

Well, it turns out that Biden was lying. Again. Two of his top military leaders—General Kenneth "Frank" McKenzie Jr., head of U.S. Central Command, and General Mark Milley, the chairman of the Joint Chiefs of Staff, testified before Congress that they had recommended to the president that the U.S. keep 2,500 troops in Afghanistan after the August 31 withdrawal deadline.

This contradicts what Mush told ABC News in August when asked if his military brass advised him to leave 2,500 troops in Afghanistan.

"No," Mush replied. "No one said that to me that I can recall."

As for Americans left behind, Mush oddly decided it was a good time to make a joke when asked by NBC's Peter Alexander, "If Americans are still in Afghanistan after the deadline, what will you do?"

"You'll be the first person I call," Biden oddly responded with a chuckle.

He took no further questions, while the White House cut his audio feed.

It's the 1970s all over again, except things overall are actually much worse. Unlike with Vietnam after the fall of Saigon, the U.S. may be forced to send our troops back in. After all, if Al Qaeda 2.0 can reorganize in the Taliban's Afghanistan, it could restore its status as a hotbed for terrorism and carry out another devastating attack on United States soil. Our failure in Afghanistan also waved a white flag at other bad actors around the globe. As all of this was happening, two of our biggest adversaries, China and Russia, were watching the president's actions—or inactions—closely. They were undoubtedly thinking, if a ragtag Taliban military could push around the mighty U.S., and American leadership was *this inept* in handling this withdrawal, this may be our opportunity, too. Especially with Mush running the show. This isn't a matter of bad luck or him being passive, either. Robert Gates, who served as Defense Secretary in the Obama-Biden administration, saw it up close and personal: Biden's instincts on foreign policy are *always* wrong. The problem now: He's the one in charge.

On February 24, 2022, Russia invaded Ukraine, marking the first land war in Europe since World War II. A Harvard-Harris poll taken shortly thereafter found that 62 percent of Americans believed Russian president Vladimir Putin would not be moving against Ukraine if Trump had been president. Logic agrees. Putin could have moved on Ukraine anytime between 2017 and 2021, when Trump was in office. He didn't. Why is that?

And at a time when the U.S. president needed to be disciplined and judicious with his words, the commander in chief wasn't up to the task.

In March 2022, when Biden went to Belgium and Poland to meet with NATO and U.S. troops, he set off several alarms, to the point that his own staff had to quickly issue clarifications and cleanups on Aisle 5.

Here's Biden (without the help of a teleprompter) on March 25 addressing U.S. troops in Poland:

> You're going to see *when you're there* [in Ukraine], and some of you have been there, you're gonna see—you're gonna see women, young people standing in the middle in front of a damned tank just saying, "I'm not leaving, I'm holding my ground."

Come again? Did Biden say that at some point U.S. ground troops would be going into Ukraine to battle the Russians? Wouldn't that possibly spur a nuclear response from Putin?

"The president has been clear we are not sending U.S. troops to Ukraine and there is no change in that position," a White House spokesperson later clarified.

The day before, during a press conference in Brussels, Belgium, Biden was asked what would happen if Russia used chemical weapons in Ukraine.

Reporter: On chemical weapons: If chemical weapons were used in Ukraine, would that trigger a military response from NATO?

Biden: It would trigger a response in kind, whether or not—you're asking whether NATO would cross; we'd make that decision at the time.

A response in kind? The U.S. would use chemical weapons against Russia if the Russians used chemical weapons against Ukraine?

"The United States has no intention of using chemical weapons, period, under any circumstance," National Security Advisor Jake Sullivan clarified to reporters later that day in another visit to Aisle 5.

Then there's Biden speaking off the cuff during a speech in Warsaw, Poland, on March 27: "For God's sake, this man [Putin] cannot remain in power," he declared.

Was this a gaffe? Or did Biden really mean it? If it was a gaffe, it was a bad one. If he meant it, that might be even worse, because regime change in a large (nuclear) country could have severe consequences for our national interest.

Biden was asked about his Putin "cannot remain in power" remark several times while taking questions from reporters, first from NBC's Kelly O'Donnell.

O'Donnell: Do you believe what you said—that Putin can't remain in power? Or do you now regret saying that? Because your government has been trying to walk that back. Did your words complicate matters?

Biden: Number one, I'm not walking anything back. The fact of the matter is I was expressing the moral outrage I felt toward the way Putin is dealing, and the actions of this man—just—just the brutality of it.

The president added that he wasn't "articulating a policy change." Biden read part of his answers directly from a note card helpfully titled "Tough Putin Q&A Talking Points," according to the European Pressphoto Agency. Talk about inspiring confidence.

Kremlin spokesman Dmitry Peskov responded that Biden's statement was "certainly alarming" and that the Kremlin would "continue to track" statements from the U.S. president, according to Reuters.

"This speech—and the passages which concern Russia—is astounding, to use polite words," Peskov added.

Americans were paying attention back home, and many were horrified by what they heard. An NBC News poll released after Biden's remarks found an overwhelming number of Americans, 82 percent, were concerned that Russia-Ukraine would result in the use of nuclear weapons, while nearly three in four said they feared the U.S. military would end up fighting in Ukraine. Biden's telling U.S. troops that they will be in Ukraine one day, his saying that we'll use chemical weapons, and his appearing to argue for Russian regime change only intensified those fears.

The same poll also found that *7 in 10 Americans* didn't have confidence in the president's "ability to deal with Russia's invasion of Ukraine."

Maybe those who voted for Biden because of his experience, particularly in the foreign policy realm, should have listened to a respected military guy like Robert Gates.

"When he was a senator—a very new senator—[Biden] voted against the aid package for South Vietnam, and that was part of the deal when we pulled out of South Vietnam to try and help them survive," Gates told NPR in 2014. "He said that when the Shah fell in Iran in 1979 that that was a step forward for progress toward human rights in Iran. He opposed virtually every element of President Reagan's defense buildup. He voted against the B-1, the B-2, the MX, and so on. He voted against the first Gulf War.

So on a number of these major issues, I just frankly, over a long period of time, felt that he had been wrong."

On domestic policy, as it pertains to the economy and inflation, Joe Biden was also showing the American people that his brain was mush on these two fronts, too.

4

COVID RULES ARE FOR ORDINARY PEOPLE

The date was April 23, 2021. U.S. president Joe Biden convened forty world leaders in a virtual summit on climate change, including Russia's Vladimir Putin, China's Xi Jinping, Britain's Boris Johnson, and Israel's Benjamin Netanyahu.

Friends, enemies. Didn't matter. Biden had made tackling climate change one of his top priorities. Now it was showtime.

The president joined the video call from the White House in what looked like the opening to *The Brady Bunch* on steroids. Except in this version, one person particularly stood apart:

The U.S. president.

Triple-vaxxed. And yet wearing a mask. Apparently the president was confused regarding the difference between a *computer virus* and *coronavirus*, because the latter can't travel through high-speed internet. Yes, Biden wasn't completely alone in his physical space during this virtual summit, but given everyone around the president is vaxxed, was this show really necessary given the extremely low risk?

Vice President Kamala Harris and climate czar John Kerry also joined the call and were wearing masks, while most other world leaders were maskless. Because those leaders don't feel this perpetual need to virtue-signal.

Here's the thing, folks—to put it as our folksy commander in chief would—as much as Joe puts on a man-of-the-people tone, he's really just a virtue-signaling elite. He's a Swamp creature through and through. The way he wears a mask like a fashion accessory, signaling his seriousness to the base, even as he seems to dismiss the actual science of aerosol transmission, shows that this is not a serious blue-collar guy trying to fix a problem. Instead, he's a privileged aristocrat, either comically overperforming COVID theater or disregarding it thanks to his privilege as president.

But even though this is all cynicism, right down to the bone, Biden *still* uses COVID as a rhetorical crutch to distract from the rest of his administration's dumpster-fire policies.

This would be the playbook for Biden throughout the first fourteen months of his presidency: *Be seen with a mask.*

Alone with the first lady on a Delaware beach? *Wear a mask.*

Walking alone on the White House lawn? *Wear a mask.*

But then all logic begins to break down: Biden will wear a mask when walking out to Marine One upon leaving the White House—only to lower said mask when yelling within one foot of reporters. He also wore a mask upon entering a Washington, DC, restaurant back in the fall of 2021, but took it off when out of sight of cameras.

From the *New York Post* on October 18, 2021:

> The Bidens were caught on camera leaving Fiola Mare in Georgetown with their masked-up Secret Service agents in tow, according to videos posted on social media. Biden could be seen carrying his mask in both hands as he left the pricey seafood establishment overlooking the Potomac River. His wife did not appear to be carrying her mask as she left.

Washington mayor Muriel Bowser, a staunch mask advocate, must have been *really steamed* at the president for breaking DC's indoor mask mandate that was in place at the time. Except she wasn't mad at all, because *she* had broken an indoor mask mandate two months earlier at former president Barack Obama's indoor birthday party back on Martha's Vineyard, along with hundreds of others who attended the gala event.

Surely the media was going to put aside its overwhelming bias in calling out Obama and other Democratic elites for their COVID hypocrisy, right? Enter the *New York Times*, which sent a reporter to the Vineyard in August 2021 to see what residents on the Massachusetts island thought about the forty-fourth president holding a party—while the Delta variant was surging, mind you—in an enclosed tent that appeared to be the size of the Astrodome.

The coverage was . . . generous:

> For his part, Mr. Obama was soliciting advice from his most trusted advisers about what the best path forward was for revamping a costly, logistically complicated party that had been months in the making. The former president had baseball caps made for the occasion that read "44 at 60." He had also hired a "Covid coordinator," or compliance officer, to ensure the safety of guests. . . .

. . . the "report" reads. Ah, a COVID coordinator. That should totally prevent a contagious virus, already being transmitted at a high level, from spreading in a closed tent with poor ventilation.

> It soon emerged that Jay-Z and Beyoncé were still invited, along with some of Mr. Obama's oldest friends from Hawaii, who were spotted Thursday and Friday bowling and golfing with the former president, who was wearing his birthday cap.

> Tom Hanks, one of the earliest celebrities to reveal he had the coronavirus last year and a longtime Obama ally, was also still invited, although it was not clear he would attend and he had recently been spotted in Greece.

Uh-huh . . .

"He has 20-plus acres of land, and everyone was going to be outside," one resident told the *Times* reporter. "You're dealing with a sophisticated crowd. I think the concerns were a bit overblown."

The whole "overblown" sentiment would mark the final paragraph in the story and therefore the last word on this controversy.

Let's unpack this: Well, for starters, this was an *indoor* party. The *Times* kept insisting it was outdoors. And when exactly did COVID not spread to those who are *sophisticated*? That's one smart virus. And apparently one that cares about social status. Yeah, boy, COVID-19 has broken out the who's who—or in today's parlance, is scrolling Twitter and passing over any user with a "verified user" blue check mark beside their name. They're the good people, you see.

Sometimes making a point on the utter hypocrisy of our media is too easy. Compare and contrast the Obama coverage with his successor, Donald Trump.

Take the *New York Times* during the 2020 Republican National Convention. "Halfway through the Republican convention, there have been some notable omissions: a traditional party platform, any mention of impeachment and a meaningful reckoning with the number of Americans who have died from the coronavirus," it reads. After this brief focus on policy, the *Times* did what it does best—focus on the "signals" its class of readers care about. Never mind a political platform . . .

> [E]ven more conspicuously absent were masks—whether on President Trump or on those in his immediate orbit, some of whom have been featured in segments that were pretaped at the White House. No masks when Mr. Trump recognized frontline workers. No masks when Mr. Trump presided over a naturalization ceremony. No masks during the made-for-TV pardoning of a former inmate. And when Melania Trump, the first lady, gave a keynote speech on Tuesday night in front of a crowd in the White House Rose Garden, very few members of the audience wore masks. The attendees, who the White House said had been tested for the virus, also did not practice social distancing.

Now, remember: The Republican National Convention in 2020, unlike ObamaPalooza, was largely held outdoors, either in the Rose Garden for the first lady's speech or the spacious South Lawn for President Trump's remarks. Regardless, it's a marvel when comparing the tone of each story. Final score: Obama gets a pass while Trump gets spanked from the alleged "paper of record."

We've seen this all too often now over the past year: Our so-called leaders—in the most pious fashion in interviews and at press conferences—declaring that we all need to do our part to slow the spread, mask up, and avoid large gatherings during this pandemic . . . only to turn around and not follow their own rules in applying the *Rules for thee, but not for Ds* (as in Democrats) mantra.

As for Mayor Bowser, she would proceed to announce in the fall of 2021 that masks had to be worn again in indoor settings in DC after the CDC deemed the city an area of substantial and high transmission of COVID-19.

In a related story, there were twenty-one homicides in July 2021 in Washington, DC, a city with a population of 700,000, compared to just *eight* COVID-related deaths. Also, twenty-four hours after announcing the renewed mask mandate, Bowser officiated a wedding at a five-star hotel before hitting the dance floor. Without a mask, of course.

"We all know what the rule says about sitting at a dining table and dining. Don't be ridiculous," Bowser said when called out on her hypocrisy. "They took a picture of me where dinner and drinks were served."

But video showed Bowser at a table with no mask after the meal portion was over, and Bowser still denied she broke any mandate. No matter: The story went away quickly.

This pathetic movie played out in other Democratic-led cities and states across the country. Democratic California governor Gavin Newsom dined indoors with twelve people without a mask in pre-vaccine 2020 at a ritzy restaurant in wine country. In 2021, the same governor pulled his kids out of summer camp over its mask policy.

Democratic California congresswoman Nancy Pelosi had her hair done indoors in San Francisco in pre-vaccine 2020—then blamed the salon owner for setting her up through some kind of a Jedi mind trick compelling her to go.

"As it turns out, it was a setup. So I take responsibility for falling for a setup, and that's all I'm going to say on that," the House Speaker explained.

And that settled it. The press moved on.

The list goes on and on. Chicago's mayor, Lori Lightfoot, also got her hair done while salons were closed under *her* rules. Democratic socialist congresswoman Alexandria Ocasio-Cortez attended a ritzy gala at the Metropolitan Museum of Art in New York City without wearing a mask (while the staff was forced to wear them). San Francisco mayor London Breed danced maskless at a live concert and got caught doing so on camera.

Breed's explanation is about as pious and contradictory as it comes. "We don't need the fun police to come in and try and micromanage and tell us what we should or shouldn't be doing. We know what we need to do to protect ourselves," she told reporters in September 2021.

The hubris of these example-setters is incredible.

Michigan governor Gretchen Whitmer implored residents of her state not to go to the (free) state of Florida, only to fly there herself. And on a private jet, no less. Her health commissioner also decided a trip to the Gulf was perfectly safe in traveling to a ritzy resort in Alabama, as did her chief operating officer, who went to the Florida Gulf Coast and was arrogant enough to post maskless party photos in the process.

"What directors do on their personal time is their business," Whitmer said. "So long as they are safe, which is what we're asking everyone in this state do to: get vaccinated, mask up."

Los Angeles mayor Eric Garcetti, along with Governor Newsom and every mask-loving celebrity in Tinseltown, attended the Super Bowl in February 2022 maskless.

In terms of explanations around breaking their own rule, you gotta hand it to Garcetti, who had easily the best response to a question since Bill Clinton's "I didn't inhale" answer on smoking weed.

"I wore my mask the entire game. When people ask for a photograph, I *hold my breath* and I put it here and people can see that," Garcetti explained while holding his mask in his hand. "There is a zero percent chance of infection from that."

If you just threw this book across the room, the author completely understands your frustration. Do these people think we're *this* stupid?

Outside SoFi Stadium, the Los Angeles Youth Orchestra performed while wearing masks. Inside, seventy thousand people, led by the governor and mayor, were maskless in front of 110 million watching at home.

And they didn't care.

Another way in which the class divide between elites and ordinary Americans manifested was in the policies we put in place for children. Rich, urban, or left-wing Americans are all far less likely to have children than poor, rural, or right-wing Americans. This may explain why elites find it so easy to put policies in place that profoundly harm childhood development, out of fear of a disease that primarily affects adults. The *New York Times*' David Leonhardt expressed in January 2022 what many had been saying for many months in an essay titled "No Way to Grow Up." He wrote, "For the past two years, Americans have accepted more harm to children in exchange for less harm to adults."

One thousand percent correct. Schools have been proven, in study after study, to be one of the safest places anyone could be.

My son, a kindergartner, hadn't been in school in his entire existence without a mask until March 2022. Virtual learning was a disaster. Suicides skyrocketed among young people during COVID.

According to the CDC's *Morbidity and Mortality Weekly Report*, trips to the emergency room shot up by 22 percent for potential suicides for kids aged twelve to seventeen in the summer of 2020 compared to the same season in 2019. Things got worse in the winter of 2020 compared to 2019, with visits for the same reason up nearly 40 percent.

The American Federation of Teachers (AFT) and CDC didn't care. In fact, they were working in concert with each other to ensure that schools stayed closed longer than they ever should have.

Unions were pulling the strings of the CDC all along.

Emails obtained in March 2022 by a group called Americans

for Public Trust, which consists of lawyers and researchers, showed that the AFT and the National Education Association received a copy of CDC guidance on school reopening *before* it was released to the public, and that the CDC allowed the unions to apply line-by-line edits to key parts of the reopening guidelines.

Needless to say, this practice was not remotely common. The CDC usually keeps its draft guidance confidential. But here, teachers union bosses *were dictating the science.*

So why do our so-called leaders act this way? Because they *know* they'll ultimately get away with it. They only care about pleasing their own class of people.

More important, how have Biden and the Dems performed on defeating COVID, which they promised they would do? This was supposed to be the relatively easier part, considering what they were handed: multiple vaccines, effective therapeutics, and a year to study the virus (and public sentiment) after it was thrust upon the country the previous March.

They completely blew it anyway.

Things came to a head in October 2021 when the administration rejected a plan by COVID testing experts to send out more than 700 million rapid at-home COVID tests ahead of an expected holiday surge. That surge came in the form of another variant called Omicron, which sounds like a marketing firm on Madison Avenue.

What separated Omicron from other variants was how highly contagious it was, with experts comparing it to chicken pox in this regard. By Christmas Eve, the country, particularly the East, was overwhelmed by the virus.

My wife, an ER physician, contacted me upon arriving at work at the time to report back what she was seeing.

"There's a line going all the way down the block. Never seen anything like this before," she texted. I would personally see the same outside of two urgent cares near our home in New Jersey: lines of people in the cold looking like those in line for bread in third-world countries. In some parts of the country, these lines would extend for miles. Case numbers shattered records.

But how could this be? Candidate Joe Biden, the guy who promised to cure cancer, said he had a plan to shut down and control the virus. As we all know now, vaccine mandates were never about science or health or safety. The Canadian truck convoy protest over vaccine mandates in Ottawa in early 2022 was a prime example, when Canadian prime minister Justin Trudeau portrayed those protesters as fascists.

"This is a story of a country that got through this pandemic by being united," he said. "And a few people shouting and waving swastikas does not define who Canadians are."

The protests themselves, unlike the mostly peaceful riots of 2020 in the U.S., *were* peaceful. Buildings weren't burned to the ground. Police officers weren't attacked. And no, there weren't armies of Nazis running around Ottawa.

But matters got profoundly ridiculous when Canadian Parliament Conservative Party member Melissa Lantsman criticized Trudeau for evoking emergency powers during the protest, giving him unprecedented power. Trudeau's response was a classic example of political goalpost shifting. He looked at a genuine grassroots, working-class protest and said it was about Naziism.

"Conservative Party members can stand with people who wave swastikas, they can stand with people who wave the Confederate flag, we will stand with Canadians," Trudeau said in response, while he surely got a standing ovation at MSNBC and at the White House.

In a related story, Lantsman is (checks notes) Jewish, making her just the type of person to stand with people who wave swastika flags.

Making this all the more ridiculous, from a science perspective, the Canadian population is one of the most vaccinated in the world, with more than 80 percent of citizens receiving two jabs. So why again is a vaccine mandate needed here for a profession where the worker is almost always isolated from the public?

This is not about science. It's about *political science.*

Here's a tweet from Ottawa police in February 2022: "All media who are attending the area, please keep a distance and stay out of police operations for your safety. Anyone found within areas undergoing enforcement may be subject to arrest."

Threatening to arrest members of the free press covering protests is a great look, right?

But for the most batshit crazy part of the COVID saga as it pertains to vaccine mandates, look no further than basketball player Kyrie Irving. Irving's résumé is beyond impressive: At Duke, a national champion. Joined the Cleveland Cavaliers as a number one pick, Rookie of the Year. 2016 NBA World Champion. Seven-time All-Star. Olympic gold medalist. I specifically remember seeing Irving play in Jersey on the high school level and had no doubt he would not only make it to the NBA, but also become a big star.

As you may have heard, Irving, who currently plays for the Brooklyn Nets, is against getting the vaccine. Here's his reasoning in October 2021:

> It is reality that in order to be in New York City, in order to be on a team, I have to be vaccinated. I chose to be unvaccinated, and that was my choice, and I would ask you all to just respect that choice. I am going to just continue to stay in shape, be ready to play, be ready to rock out with my teammates and just be part of this whole thing. This is not a political thing; this is

not about the NBA, not about any organization. This is about my life and what I am choosing to do.

Despite what you may have read or seen in the media about Trump supporters leading the charge on resisting the vaccine, that hesitancy is actually highest among the Black community. In October 2021, at the time Irving shared his perspective before the start of the season, only about 40 percent of the Black population was vaccinated. For context, more than 70 percent of the Asian population and 50 percent of the white population were vaccinated at the time.

But political leaders in New York City, which forced kids to eat outside in freezing temperatures during the winter of 2021–22 while eighteen thousand people could scream at a Knicks or Nets game jammed together without masks, decreed that Irving could not play in any home games as long as he stays unvaccinated. And even after Irving got COVID (Omicron) in December 2021, the city still wouldn't consider the natural antibodies he had.

There was one Sunday afternoon in March 2022 when I nearly fell off my couch after witnessing just about the stupidest thing you'll ever see when it comes to COVID and "science."

The network: ABC. The game: New York Knicks at Brooklyn Nets. Kyrie Irving wasn't allowed to play for the home team. But there he was, sitting courtside like a young Spike Lee, cheering his team on, without a mask, and even greeting them during time-outs with high fives and chest bumps.

Can you believe this? Irving is allowed to sit in the stands not wearing a mask. He can touch his teammates or talk to people in the stands at close range while maskless. But *going on the court and playing* was deemed a health risk. Players obviously aren't masked out there, but the officiating crew is.

Yep. Makes total sense.

Oh, and here's the best part: If a *visiting player* from another state is unvaccinated, the rules in New York say it's perfectly fine that he plays.

Because, you know, *science*!

New York mayor Eric Adams finally allowed Irving and other unvaccinated professional players in other sports to participate in home games. But in the ultimate insult, the new mayor did not extend that "courtesy" to first responders and medical workers, who only put their lives on the line throughout the pandemic before getting canned.

"My body, my choice" suddenly didn't apply.

Here's the bottom line: Without COVID, Joe Biden isn't president. It provided an excuse for him to avoid campaigning, for him not to be exposed to those in the press who cared to do their job correctly. COVID was the dominant issue in the 2020 campaign, and Biden exploited it to the fullest by apparently convincing enough people to believe that he could stop and control it.

Going into 2020, Donald Trump had achieved the two basic things that matter in a first-term presidency: Our economy was strong and we were largely at peace. People were secure in their jobs and with their 401(k)s. The last two one-term presidents, Bush 41 and Carter, both faced recessions in an election year. Trump's economy was quite the opposite.

Put Trump's rhetoric and mean tweets aside, because in a normal, non-COVID circumstance, Biden would have actually been required to articulate an argument across the country and in face-to-face interviews that his economic policy (spend more!) and border policy (open it up!) and education policy (teachers unions know better than parents!) and foreign policy positions (see: Mush) would make America a better place than Trump already had achieved.

And given Biden's inability to make such a case without the full aid of a teleprompter and index cards, it's a great bet Trump would have won, because ultimately people vote with their wallets while feeling safe in their communities.

But COVID did come, thanks to the Chinese. Something else that came was inflation, largely thanks to reckless spending. And with inflation came the comparisons to Jimmy Carter for Joe Biden. That isn't a good thing.

5

EVERYTHING IS FREE IF YOU WISH HARD ENOUGH

The forty-sixth president of the United States walked to the podium. His mask, of course, was on, despite his being fully vaccinated and COVID cases being exceedingly low at the time.

It was the fall of 2021. Joe Biden was polling in the 30s among independents and dropping quickly. His handling of the U.S. drawdown from Afghanistan would mark the beginning of the end of his presidency, at least in terms of being trusted as a competent leader by a majority of the American people.

Unless . . .

A Democratic bill called "Build Back Better" could be rammed through the Democrat-controlled House and the 50–50 Senate (with Kamala Harris as the tiebreaker) for his signature. No Republicans would be needed, so the president and his party didn't even attempt to bring them on board.

So much for that whole unity thing. Looking back, one could easily argue that Biden (or more likely his handlers like Chief of Staff Ron Klain) pushed all the chips to the middle of the table and bet his presidency on this monstrosity of a bill passing. How big a monstrosity? Try $5.5 *trillion* in new spending after stripping away the budget-accounting gimmicks. If passed, BBB would result in the largest expansion of the U.S. government in our nation's history.

So what was in this bill?

Well, apparently building back better includes:

$79 billion for the Internal Revenue Service to expand tax enforcement (because voters *love* more IRS in their lives)
$3 billion for a tree-planting program that increases "tree equity"
$1 billion for an "electric vehicle charging equity program"
$7.5 billion for the launch of the "Civilian Climate Corps"
$7 billion to the U.S. Postal Service to convert all their vehicles to electric power
$45 billion for free community college
$110 billion for universal prekindergarten learning

Remember, this bill came on top of the $2 trillion "COVID relief" bill passed earlier in his presidency in March 2021, which included many items that didn't have anything to do with, you know . . . COVID. Here are a few items from that disaster.

1. "Of the funds appropriated under title III of the Act that are made available for assistance for Pakistan, not less than $15,000,000 shall be made available for democracy programs and not less than $10,000,000 shall be made available for gender programs."

   You read that correctly. $10 million.
   For gender programs.
   In Pakistan.

2. Funds for "Resource Study of Springfield (Illinois) Race Riot."

   That riot occurred in (checks notes . . .) 1908.

3. "Statement of Policy Regarding the Succession or Reincarnation of the Dalai Lama."

Big hitter, the Lama. We'll just leave that one there.

4. A commission tasked with educating "consumers about the dangers associated with using or storing portable fuel containers for flammable liquids near an open flame."

Yep. An entire commission.

5. Another $40 million allocated "for the necessary expenses for the operation, maintenance and security" of the Kennedy Center.

Throw in $86 million for assistance to Cambodia, $130 million to Nepal, $135 million to Burma, and $700 million to Sudan . . . all while American cities are crumbling as drug use and homelessness explode.

The bill also created a Women's History Museum and an American Latino Museum as part of the Smithsonian Institution. Overall, the Smithsonian was to receive (checks notes again) $1 *billion.*

In the case of Build Back Better, it might as well have been called Built Back Broke, because this was, just like COVID relief, the Porky Pig of spending bills. But our Democratic leaders—especially the president—insisted it wouldn't cost almost all taxpayers one penny. Biden and his allies also insisted that BBB would actually reduce the deficit while lowering inflation, which is totally how basic economics and math work, right?

So when the president tried to sell this thing the way tobacco executives once tried to sell the country back in the 1990s that nicotine isn't addictive, he decided to improvise in quite possibly the creepiest, most insulting way possible.

"We talk about price tags," Biden said to reporters in September 2021 as he leaned forward into the microphone with his eyes wide.

"It is a zero price tag on the debt," he claimed while speaking in a loud whisper that could only be compared to that ghost in *The Amityville Horror* telling a terrified family to "GET OUT."

"We're paying—we're going to pay for everything we spend," the whisperer in chief assured everyone. The president would go on to do the whisper thing on multiple occasions despite not being received well by most sane and sober observers (#CreepyJoe would trend on Twitter whenever the whisper reemerged).

What's so bad about it? For one thing, it's just a bizarrely weird tic. But for another, it feels patronizing, faux-authentic. "I'm letting you in on a secret," Biden implies, before explaining why he's right about everything and we don't need to ask questions about it. The government knows best. Don't ask about anything like the *facts*. As he might say: Come on, man.

On cue, of course some in the media embraced it, most notably the Associated Press, which *totally* would have reacted the same way if Donald Trump acted in such a bizarre fashion.

"Listen Up: Biden Speaks Volumes in a Whisper to Make a Point," reads the AP headline to what was somehow not an opinion piece, but an actual straight news story. Check out this excerpt:

> Biden folded his arms, rested on the lectern, leaned into the mic and lowered his voice. "Hey, guys, I think it's time to give ordinary people a tax break," he said, almost whispering

as he addressed his critics. "The wealthy are doing fine."
It was the latest instance of Biden speaking volumes by whispering. The White House and communications experts say Biden's whispering is just this veteran politician's old-school way of trying to make a connection while emphasizing a point.

If you ever needed that last crumb of confirmation that the media is hopelessly biased, this was a very big one.

Jen Psaki shared that condescending attitude when she was asked about how increased government spending will lower inflation.

"No economist out there is projecting that [the Build Back Better bill] will have a negative impact on inflation," she responded in a way a schoolteacher tells a second grader that chalk isn't edible. "And actually, what it will help do is it will help increase economic productivity. It will help economic growth in this country. That, and the Build Back Better Agenda will help reduce inflation, will help cut costs for the American people over the long term."

But many economists—and anyone with the business acumen of a kid running a lemonade stand—disagreed with Psaki, most notably Steven Rattner, who served as counselor to the Treasury secretary in the Obama-Biden administration.

"For the Biden administration, which has long insisted that prices would rise far more slowly, inflation is now its biggest challenge," Rattner wrote in a *New York Times* op-ed in November 2021.

"We worried that shoveling an unprecedented amount of spending into an economy already on the road to recovery would mean too much money chasing too few goods," Rattner, now an MSNBC analyst, added. And he's right because basic common sense is being applied: Too much money flooding the system. Too few goods to keep up with the artificial pumping.

So this isn't about math or economics or sane fiscal policy, but about satisfying AOC and the Squad wing of the party. Listening to the likes of Ocasio-Cortez, an economics major at Boston University before becoming the most popular lawmaker on social media, is *never* a good idea.

Here, in a 2018 PBS interview, is AOC's take on why unemployment was so low under President Trump. "Unemployment is low because everyone has two jobs. Unemployment is low because people are working 60, 70, 80 hours a week and can barely feed their family."

Got that? If an individual works two jobs, he or she counts *as two people*, thereby lowering the overall unemployment number.

Yep. Let's listen to *her*.

Fortunately, two lawmakers who didn't listen to Biden or AOC were Senators Joe Manchin (D-WV) and Kyrsten Sinema (D-AZ). Despite overwhelming pressure, both publicly rebuked their fellow Democrats, particularly regarding their support for abolishing the filibuster.

The president was so unhappy with this opposition, he decided to respond while speaking in the third person.

"I hear all the folks on TV saying, 'Why doesn't Biden get this done?'" Biden moped in a speech in 2021 as Build Back Better was struggling to gain Manchin's or Sinema's support. "Well, because Biden only has a majority of effectively four votes in the House, and a tie in the Senate, with two members of the Senate who vote more with my Republican friends."

Not surprisingly, this is not true. According to FiveThirty Eight.com, Manchin and Sinema had voted with Biden's position *100 percent of the time* to that point. For context, Senators Elizabeth

Warren (D-MA) and Bernie Sanders (I-VT) have each voted with Biden's position 97 percent of the time.

Democratic activists in the media also demanded that the filibuster be abolished, warning that not getting rid of it would result in everything from racism at the polls to a Democratic wipeout in the 2022 midterms. The filibuster, of course, requires that a bill be passed with at least 60 votes in the 100-member Senate. By abolishing it, a party in power, like the Democrats in 2021 and 2022, can ram through whatever they like if completely unified. But if the filibuster stays, Democrats would need to (gasp!) compromise with the opposition party.

"In a Senate without a filibuster, Democrats have some chance of passing some rough facsimile of the agenda they've promised," the *New York Times*' Ezra Klein also stated in a January 21, 2021, piece. "In a Senate with a filibuster, they do not."

"They have preferred the false peace of decorum to the true progress of democracy. If they choose that path again, they will lose their majority in 2022, and they will deserve it," Klein added.

Uh-huh. Nothing screams democracy like one party ramming through legislation based on lies around cost and inflationary and deficit impact, all while blowing up the filibuster.

Speaking of the filibuster, here's the 2005 version of Biden fervently defending it.

"It is not only a bad idea [to abolish the filibuster], it upsets the constitutional design and it disservices the country," Biden said at the time. "No longer would the Senate be that 'different kind of legislative body' that the Founders intended. No longer would the Senate be the 'saucer' to cool the passions of the immediate majority."

Manchin and Sinema, channeling Biden circa 2005, have stated that they will not back abolishing the filibuster, which the current software update of Biden—billed as a unifier—now supports doing.

House Speaker Nancy Pelosi admonished the press in October 2021 for not doing a better job of selling Build Back Better. "You all could do a better job of selling it, to be very frank with you. Because every time I come here, I go through the list, family medical leave, climate, the issues that are in there," she argued without broaching cost.

Again, picture Trump speaking this way to the press. "You should be selling this on my behalf better!"

Then picture the media reaction.

It was Pelosi who said this, however, and therefore the outrage machine from the press was turned off as a result. This is the same person who said in 2010 regarding another massive stimulus bill under Obama-Biden: "We have to pass the bill so that you can find out what is in it."

But Manchin and Sinema continued to stand firm despite shameful bullying tactics by members of the party of tolerance. Exhibit A of this behavior occurred when activists harassed and filmed Sinema while she was using a bathroom at Arizona State University, where she teaches a course.

Filming someone in a bathroom in Arizona is a Class 5 felony, but here's the way one Bloomberg reporter described the harassment on Twitter:

> "Protesters followed Senator Sinema into the bathroom at Arizona State University to confront her on Build Back Better and immigration"

And check out the passive framing from other outlets:

Daily Beast: "Senator Kyrsten Sinema locks herself in bathroom to avoid young activists on ASU campus"
New York magazine: "The Ugly Truth About the Kyrsten Sinema Bathroom Protest"

Let's recap. The Biden administration was whispering its extravagant budget to an eager mediasphere. While Obama-era economists like Rattner and Larry Summers attempted to pump the brakes, Team Biden was far more interested in staying in AOC's good graces. So they relied on that classic liberal lie: Everything can be free if you wish hard enough. Never mind things like debt or scarcity or supply and demand . . . the world really is full of unicorns farting rainbows. Or something.

Maybe this is why the media was so desperate to defend Sinema's persecutors. "Aggressive protest tactics like the one that so incensed Sinema tend to be a symptom of a much larger problem. When people are shut out of a supposedly democratic process, they have no choice but to agitate," *New York*'s Sarah Jones argued in her "Ugly Truth" screed in October 2021.

Think about the argument here: "They have no choice but to agitate."

This kind of rhetoric comes from people like Jones despite what happened not too long ago during a congressional baseball practice outside of Washington, DC, when a progressive maniac opened fire on Republican lawmakers in 2018, wounding five, including Representative Steve Scalise (R-LA), who barely survived. Encouraging "protesters" to directly "agitate" lawmakers is the height of journalistic irresponsibility. It is absolutely certain that Jones would never advocate this kind of behavior if say, pro-life advocates followed Alexandria Ocasio-Cortez into a bathroom at New York University and harassed *her*.

In Sinema's case, the person who filmed her illegally was on Facebook bragging about the actions of her group. Facebook, which would have likely pulled her post in about three seconds if a progressive member of Congress had been harassed in the same manner, allowed the video to stay up, as did Twitter. Talk about comically selective "enforcement" of their own rules . . .

Joe Biden should have condemned this harassment of a female senator, but instead he argued that this sort of thing happens to everybody and is "part of the process."

"I don't think they're appropriate tactics, but it happens to everybody," Biden said at the time. "The only people it doesn't happen to are the people who have Secret Service standing around them. . . . So, it's part of the process."

No, Mr. President. Chasing a lawmaker into the ladies' room with a camera and recording her is not "part of the process."

Biden had an opportunity to bring down the temperature here, to unify, but once again he found a way to show that the compassion he campaigned on was just empty talk.

By the way, it wasn't that long ago that bucking your party meant being given icon status, as witnessed by the media's treatment of John McCain after he voted against President Trump and the Republican Party's efforts to repeal and replace Obamacare in 2017.

McCain headlines:

Washington Post: "Analysis: The Iconic Thumbs-Down Vote That Summed Up John McCain's Career"

New York Times: "McCain Provides a Dramatic Finale on Health Care: Thumb Down"

CNN: "The 'Thumbs Down' Health Care Vote That Enraged Trump Is John McCain's Lasting Legacy"

Iconic. Dramatic. Lasting legacy.

Contrast this with Sinema's headlines:

Washington Post: "Despite Manchin and Sinema, Democrats Are More United Than They've Been for Decades"

New York Times: "Kyrsten Sinema Is at the Center of It All. Some Arizonans Wish She Weren't."

CNN: "Sinema Won't Commit to Voting for Biden's Sweeping Social Safety Net Expansion"

The reality is, you're only allowed to be a maverick if you lean left. If you lean right, you're letting down the team. You're an obstructionist.

But imagine a different world. A world where politicians wanted to achieve real change—in other words, a world where they had the humility to seek bipartisan support. The plan should have been to chalk up moderate wins that Manchin and Sinema and even some Republicans could support to get to sixty votes. The easiest win of all could have been (and could still be) lowering the cost of prescription drugs. This is one of those rare areas where red and blue voters overwhelmingly agree.

The nonpartisan Kaiser Family Foundation found that 95 percent of Democrats, 82 percent of independents, and 71 percent of Republicans support the idea of Medicare negotiating with pharmaceutical firms to lower the prices for both its beneficiaries and those who have private insurance.

Polling by the Robert Wood Johnson Foundation was similar, with 84 percent of those polled saying the government should be permitted to cap prices for drugs that can help save lives and for common chronic illnesses.

Here's why it matters: More than 1.1 million U.S. Medicare patients could die over the next ten years because they can't afford to pay for their prescription medications, according to the nonpartisan West Health Policy Center.

But for whatever reason, Biden and the Democrats stuffed this into Build Back Better and all the pork and unnecessary spending that went along with it. It's completely ridiculous and is why the president and Congress poll so poorly with the American people: They can't get the simplest shit done to help save lives.

Build Back Better would ultimately fail despite the efforts of Biden, Pelosi, and the press to push it over the finish line. In 2021 and 2022, Democrats controlled the House. They essentially controlled the Senate. Its nominee got more votes than anyone in U.S. history.

Yet they still couldn't pass their most prized agenda. However, they had still pushed through so much spending that America was about to face a historic crisis. One that "the adults in the room" should have seen coming. And maybe they would have, if they hadn't been lost in that great orchard of money trees growing in AOC's imagination . . .

6

BIDENFLATION

The date was March 11, 2022, and the president of the United States had morphed into Grandpa Simpson again.

"I'm sick of this stuff!" Joe Biden bellowed at a Democratic conference in Philadelphia.

He then gestured with his hands under his neck, as if he had it up to here. "The American people think the reason for inflation is the government spending more money! Simply not true!"

I'm not much of an economist, but I did take Econ 101. I distinctly recall learning that if you flood the system with trillions of dollars in additional spending, it could lead to higher inflation. Surely the Democrats wouldn't try to insist that government spending will both supercharge the economy *and* that supercharging the economy will not lead to inflation, would they? No matter. This is the hill Democrats have decided to die on heading into the midterms.

"When we're having this discussion, it's important to dispel some of those who say, well, it's the government spending," Nancy Pelosi said of inflation and spending at the same conference. "No, it isn't. The government spending is doing the exact reverse, reducing the national debt. It is not inflationary."

Government spending is reducing the national debt? Fascinating. Every rule of economics had been turned upside down. What should you do if you get into debt, folks? Run down to the store and buy the biggest TV you can find. Here's what the Congressional Budget Office had to say about "spending your way out of debt," according to TaxFoundation.org:

> If the legislation's other temporary tax credits and spending programs are also made permanent, the cost would approximately double to about $5 trillion and add $3 trillion to the national debt over 10 years. . . . The lack of stable funding required to extend the new programs creates uncertainty for taxpayers and beneficiaries as they may not be able to rely on the programs going forward, or new taxes will need to be levied.

So here we have Biden and Pelosi insisting that increased government spending decreases the national debt, while the CBO says it will add $3 trillion. Who do you believe here? I recommend trusting the answer that aligns with, y'know, common sense and math.

The reason for our insane spike in inflation in 2022 is government spending, of course, which ballooned in Biden's first year. Build Back Better was dead in the water, but other spending packages did pass in Biden's first year: $1.9 trillion in "COVID relief" in March 2021 (which is on top of the $2.3 trillion signed into law by President Trump when the pandemic hit in early 2020) and another $1 trillion for the bipartisan infrastructure package. Add it all up and the U.S. had injected $5.2 trillion into the system.

Right on cue, inflation rose to a forty-year high to kick off 2022. Biden should have known this would happen. As he's so fond of boasting, he's been in the political game for a long time. But if you think Joe Biden is going to take any responsibility for any of this, think again.

Hmmm . . . who do we blame this time? Biden and his handlers must have been thinking. *Manchin? Sinema? Trump? Naaaah. Let's go full* Rocky IV*!*

Enter Vladimir Putin.

"Today's inflation report is a reminder that Americans' budgets are being stretched by price increases and families are starting to feel the impacts of Putin's price hike," Biden said after the Department of Labor reported a 7.9 percent rise in consumer prices for February 2022, putting inflation at its highest level since 1982 for the third straight month.

Sorry, *Putin*'s price hike?

"A large contributor to inflation this month was an increase in gas and energy prices as markets reacted to Putin's aggressive actions," the president also said.

Yes, inflation is *Putin*'s fault. Russia invaded Ukraine in February 2022 and inflation and gas prices just shot up suddenly in the United States, right? Before that, everything was effing hunky-dory. Except even the left isn't buying this, including those who once worked in the Obama-Biden administration.

"Well, no," Obama-Biden alum Steve Rattner tweeted in response to Biden's blame-Putin gambit. "This is Biden's inflation and he needs to own it."

As gas prices hit an all-time high in March, the president was asked about where prices were headed. "They're going to go up," he responded.

When asked about solutions, he offered this response: "Can't do much right now. Russia is responsible."

No sane soul believes this, of course. Seventy percent of voters in an ABC News poll taken at the time disapproved of Biden's handling of inflation and gas prices, which goes well beyond just Republican opposition.

"I expect our Democratic friends will now try to blame the entire increase in prices on our efforts to punish Russia. But don't be fooled," Minority Leader Mitch McConnell (R-KY) said on the Senate floor. "This was more than a year in the making." Fact check: true.

When we look back on the Biden presidency, the comparisons to Jimmy Carter are impossible to avoid. As my colleague Charlie Gasparino has said, "Inflation is theft." It's a tax on everyone, with the low and middle class hit the hardest because they can't afford to absorb the increase of costs for everything from food to gas to home heating bills. Add it all up and it's 1979 all over again.

In blue states, folks are fed up. According to the Census Bureau, eighteen different states, including the District of Columbia, saw their populations *drop* from June 2020 to July 2021. The biggest decreases occurred (not surprisingly) in New York, which lost more than 310,000 people; California, which saw more than 260,000 flee; and Illinois, which experienced a drop of more than 110,000.

You'll never guess which party runs those states . . .

Meanwhile, the red state of Texas *gained* about 310,000 new residents and Florida put out the welcome mat for more than 210,000. Why not? Warmer weather. No state income taxes. Far fewer COVID restrictions. What's there not to love?

The old days used to dictate that an individual had to stay near a major city like New York or San Francisco or Chicago for work reasons. That's largely not the case now. If there's one positive thing that came out of this horrific pandemic, it's that employers finally understood that those working from home are far happier and productive than those going to the office.

I'll use myself as an example: For years, I would have to travel into the city to go on Fox News. The network was nice enough to send a car, but that still meant traveling anywhere from 60–90 minutes going into the studio for a five-minute hit (which I had to be decidedly early for), and then when it was over, it was back to the car for another 60–90 minutes.

Here's the funny part: From a hill right up the road from me, you can *see* the city skyline quite clearly, since we live only twenty miles from the Big Apple. But traffic profoundly sucks around here, and mass transit is painfully slow, so to do a five-minute hit, it used to take almost *three to four hours* out of my day.

So I got a home studio, which isn't as elaborate or expensive as it sounds and takes up little space in my basement. Getting that time back was a game changer by allowing me to appear on the network more *and* keeping my responsibilities as the kids' perpetual Uber driver to school and practices when my wife is working.

The friends I speak to specifically, at least as it pertains to New York City, all said the same thing when they left: The COVID restrictions, especially on their kids in school, were too much. The surging crime made them feel unsafe. They used to not even give it a second thought under Mayor Adams's predecessors Rudy Giuliani and Mike Bloomberg.

The homeless—often suffering from mental illness—are seemingly everywhere, they would also note. A walk down every block invariably has the scent of weed in the air; just take a stroll there sometime. Also, New York's high taxes and exorbitant cost of living made it easy to decide to move to Florida or Texas. One friend even moved to Utah (big skier) without missing a beat with his job, while another went to Kentucky, most likely for the bourbon culture.

A 2022 survey from United Van Lines showed that, besides Florida and Texas, the top inbound states that folks migrated to the previous year were Vermont, South Dakota, South Carolina, and West Virginia. In a related story, all those states have Republican governors.

The reality is, in 2022, we live in a more expensive country. A less safe country. And an easier country to enter illegally. All the spin in the world doesn't change what people feel and see on the ground.

First, inflation was deemed "transitory," according to Team Biden and the head of the Federal Reserve, Jerome Powell. Then it was about greedy corporations engaging in price gouging to line their own pockets. Joe Biden, who has never run a business in his seventy-nine years on the planet, offered a solution to inflation during the 2022 State of the Union address.

"We have a choice," he said. "One way to fight inflation is to drive down wages and make Americans poor. I think I have a better idea to fight inflation: lower your costs, not your wages."

Lower costs and increase wages? Why didn't anyone think of *this* sooner? It's a corporate utopia! Biden and Pelosi's plan, in addition to this genius advice, is to spend more money to lower inflation. Assuming you put basic math and logic aside, *this just might work.*

One of the few economists ahead of the curve was Larry Summers, who served as President Clinton's Treasury secretary and as director of the Obama-Biden National Economic Council. Again, this isn't a conservative criticizing a Democratic president here:

"We are printing money, we are creating government bonds, we are borrowing on unprecedented scales," Summers said all the way back in May 2021, when inflation was portrayed as a transitory speed bump. "Those are things that surely create more of a risk of a sharp dollar decline than we had before. And sharp dollar declines are much more likely to translate themselves into inflation than they were historically."

"I think they should learn the lesson of the Johnson administration's errors that elected Richard Nixon and the Carter administration's errors that elected Ronald Reagan," Summers also cautioned.

Summers was right. The Federal Reserve is printing money at an alarming rate. See what the *Washington Post* has to say about it (emphasis mine):

> Over the past two years, as the Federal Reserve fought to rescue the economy from the clutches of the coronavirus, *the central bank's emergency remedies increased the nation's money supply by an astonishing 40 percent.*
>
> That was almost four times as much new money as had been created during the two years that preceded the pandemic and, to some Fed critics, explains why the United States is experiencing its highest inflation since 1982.

The White House, knowing it had a losing hand here going into an election year, turned to their pals in the media in December 2021 "behind the scenes" to attempt "to reshape coverage in its favor," according to a CNN report. The conversations were described as "productive, with anchors and reporters and producers getting to talk with the officials."

Awesome! Productive conversations! And since those meetings occurred, the coverage did appear to change.

"It seems as if there is nothing that the White House can do to improve their political standing these days," lamented Chuck Todd on *Meet the Press*. "It does feel like every week there is another poll. It's a new bottom, it's a new this.

"Some of it is out of their control. His two big promises were to get COVID behind us and we get rid of Donald Trump. COVID is not behind us and Donald Trump is still lurking. It is not his fault, but is that why we are in this no man's land for him?"

That sound you hear is the late, great Tim Russert turning over in his grave.

I often talk on TV about the disconnect between the media, many of our political leaders, and regular folks watching and reading at home. This was never more obvious than during Biden's first year, when inflation was yet another issue that the White House somehow didn't see coming. When they finally did address the issue, the tone-deafness was really, *really* something.

They claimed that these were first-world problems, something so insignificant that they shrink to mere *inconveniences* in light of how much worse things could be. "Most of the economic problems we're facing (inflation, supply chains, etc.) are high class problems," Jason Furman, the former chairman of the Obama-Biden Council of Economic Advisers, tweeted. "We wouldn't have had them if the unemployment rate was still 10 percent. We would instead have had a much worse problem."

White House chief of staff Ron Klain, a prolific tweeter (and that's not a compliment), loved this argument so much that he retweeted it while also writing "This" with two finger emojis pointing down at Furman's tweet.

Yep, the inflation and supply chain crises are "high class problems." Because only the top 1 percent drive cars and purchase gasoline, correct? Only the rich buy meat, eggs, fish, and poultry. Only the fat cats heat their homes.

The same kind of elitist, tone-deaf messaging was also carried out by Jen Psaki when she was asked if the president should have moved more quickly to address the supply chain crisis.

"Ah, the tragedy of the treadmill that's delayed," Psaki responded with the usual condescension.

No, not just treadmills, Jen. Try food and medicine and medical supplies and auto and machine parts. Oh, and baby formula. If looking for the perfect way to sum up the Biden presidency on the messaging front, here you go!

Jen Psaki, May 13, 2022: "Baby formula is so specialized and so specific that you can't just use the Defense Production Act to say to a company that produces something else, 'Produce baby formula.'"

Joe Biden, May 18, 2022: "I'm taking two new steps to increase baby formula supply: Invoking the Defense Production Act to increase domestic production."

Perfect. Transportation secretary Pete Buttigieg also got into the act. "Demand is off the charts, retail sales are through the roof. . . . Demand is up because income is up, because the president has successfully guided this economy out of the terrifying recession."

In a related story, we're governed by children. Of course, demand is up because people have more cash. But since it was an artificial cash injection, the market didn't have time to make enough supply to meet demand. That's how you get a supply chain crisis.

Exhibit A is energy secretary Jennifer Granholm, former Michigan governor and CNNer, who laughed like she was watching an episode of *Veep* when asked recently what she would do to address rising gas prices.

Team Biden says she was appointed because she was governor of a state where the auto industry exists. So who better to solve the gas crisis, right? Because cars *run on gas*! It makes one believe that Hunter Biden may have been more qualified for energy secretary, given all that experience with Burisma in Ukraine and riding Amtrak and all . . .

But here's where things make zero sense. Here we have many in the Democratic Party constantly demanding the two most regressive spending decisions you'll ever see: forgive student loans (the majority of college students and borrowers are white students) *and* repeal SALT, the adjustment to the income tax deduction for State and Local Taxes instituted under Trump and which cuts taxes primarily for wealthier blue state homeowners. As of March 2022, Biden has already forgiven $17 billion in student loans.

So with his back against the wall on inflation, Biden decided to channel Grandpa Simpson again in an effort to get the press back on his side. What better way to do that than to attack a Fox News correspondent on a personal level?

"Do you think inflation is a political liability going into the midterms?" Fox News White House correspondent Peter Doocy asked in early 2022.

"No, it's a great asset," Biden muttered sarcastically in response. "More inflation. What a stupid son of a bitch."

Hey! It's that time in the book where we play our favorite game: *What if Trump said that?* Would the media reaction look like the following, from the *New York Times* after the president called a reporter a stupid son of a bitch?

"Joe Biden and Peter Doocy Is the Rivalry Everyone Can Love," read the headline. "A notably unscripted exchange between President Biden and a Fox News correspondent went viral—and turned into what these days counts as a heartwarming civic moment."

WTF? A heartwarming civic moment? Someone got paid to write *this*?

How about CNN?

They went with the "*But Trump!*" defense.

"Why Biden's 'son of a bitch' moment *is nothing like* Trump's attacks on reporters" was their takeaway. And the British newspaper the *Guardian*? "Joe Biden *appears to insult* Fox News reporter over inflation question." Appears? It's literally on tape!

Now let's compare those to headlines from the Trump era, when a similar salvo was sent from the president to a reporter.

CNN, March 2020: "Trump Viciously Attacks NBC News Reporter in Extended Rant After Being Asked for Message to Americans Worried About Coronavirus"

Washington Post, May 2020: "Trump's 'Ask China' Response to CBS's Weijia Jiang Shocked the Room—and Was Part of a Pattern" (because remember . . . holding China accountable was seen as a *big* no-no by the press at the time)

And the *Guardian*?

May 2020: "How Trump Has Berated, Insulted and Demeaned Female Reporters"

This game never gets old.

Joe Biden. Jimmy Carter. The similarities are striking. Inflation. Gas prices. An unstable world with the U.S. looking increasingly weak to our adversaries. But perhaps Biden could at least make American cities and towns safer as crime continued to soar, right? He is the author of the tough 1994 crime bill, after all.

That Biden, unfortunately, was long gone.

7

PSAKIBOMB!

Jen Psaki is a legitimate star in the verified-blue-check Twitter world. Former Fox News anchor turned CNN streamer Chris Wallace called her "one of the best press secretaries ever." It's no wonder the press loves her. She's got the same snobberies and blind spots that they do, and lives in a beautiful elite bubble where crime is simply a sinister fiction dreamed up by the dastardly GOP. The media loves pretending we live in the fantasy world Jen Psaki describes. My former home, Mediaite, was guaranteed to run a glowing article on her almost *every day*. Here's a selection culled from hundreds of examples:

Jen Psaki Destroys Ted Cruz for Claiming Biden Pledge to Nominate Black Woman is "Offensive"
WATCH: Jen Psaki Smacks Down "Hilarious" Criticism That Biden's Voting Rights Speech Was Too "Offensive"
Jen Psaki Rips Gov. Kemp to Local Reporter for Fighting Vaccine Mandates: Help Save Lives or "Get Out of the Way"
Jen Psaki Scolds Male Reporter over Abortion Question: "You've Never Faced Those Choices"

The love extended out to larger publications as well. "When Psaki first appeared in the press briefing room, in January 2021, there was a collective swoon from roughly half the country," a *Vogue* PR piece begins, in an awe-filled tone. "This was largely due to what she was *not* doing: berating the assembled reporters, griping about CNN's coverage of a presidential tweet, or spouting flagrant, easily disprovable lies."

"Crisp and precise in her answers, even if she does not always respond directly to a reporter's questions, Ms. Psaki, in her speaking style, is a contrast to Mr. Biden and his circuitous folksiness," reads the *New York Times*, in a piece sentimentally titled "Bully Pulpit No More." "In interviews, Washington correspondents often used the word 'professional'—high praise in DC—to describe interactions with her, deeming her straightforward, detail-dense briefings a relief after an era when Mr. Trump's press secretaries repeatedly insulted, denigrated and frequently ignored journalists."

Among a sea of incompetent fools, Psaki is easily the smartest kid in summer school. Liberals on social media adore her in ways not seen since Michael Avenatti, as evidenced by her hashtag #jenpsaki and #PSAKIBOMB on TikTok generating hundreds of millions of views.

In a related story, she's also a professional liar despite all this effusive praise, yet rarely got dinged by the press the way Trump's press secretaries did. At one press briefing in March 2021, she insisted that the U.S. southern border was "closed." Someone should ask the more than 2.1 million migrants who entered the country illegally in Biden's first year if that's the case, or perhaps the drug cartels pushing deadly fentanyl into the country and killing a record number of Americans could also weigh in on that claim.

When asked again why so many migrants were entering the country so easily, Psaki went back to the 2020 campaign playbook: When in doubt, even for problems this administration has created completely on its own . . . blame Trump!

"We recognize this is a big problem," Psaki said during a briefing with reporters on March 15, 2021. "The last administration left us a dismantled and unworkable system and, like any other problem, we are going to do all we can to solve it."

Ah yes, the last administration left a dismantled and unworkable system to the poor Biden administration. The ability to play victim on everything is almost admirable on some level.

To state the obvious, it was *Biden* who stopped all construction on the border wall. It was *Biden* who ended the Trump's administration's Remain in Mexico policy, by which "asylum seekers arriving at ports of entry on the U.S.-Mexico border will be returned to Mexico to wait for the duration of their U.S. immigration proceeding," according to the Latin America Working Group. It was *Biden* who tapped an uninterested and inept Kamala Harris to handle the border catastrophe, which has only gotten worse under her watch as the number of illegal border crossings continues to rise.

Psaki also said on August 24, 2021, that the U.S. withdrawal from Afghanistan couldn't be called "anything but a success," despite the obvious and accurate comparisons to the chaotic withdrawal from Saigon at the end of the Vietnam War in 1975.

Psaki also claimed on August 23, 2021, that it was "irresponsible to say 'Americans are stranded' [in Afghanistan]. They are not."

No, they were. Nearly two hundred Americans, according to the State Department in October 2021. Psaki once claimed that it was *Republicans* who supported defunding the police, which promptly earned her three Pinocchios from the *Washington Post.*

To be fair, she had a tough job spinning the Biden administration's lack of accomplishments. Once, the White House touted 16 cents in savings that Americans "enjoyed" over the Fourth of July. "I would say if you don't like hot dogs, you may not care [about] the reduction of cost," the press secretary said, lamely, scolding a reporter for even questioning the ridiculous boast.

How about crime skyrocketing in many American cities and

towns across the country? That's just fear-mongering from "alternative universes" in the media, according to Psaki.

Fox's Jeanine Pirro "is talking about 'soft-on-crime consequences.' I mean, what does that even mean, right? So, there's just an alternate universe on some coverage. What's scary about it is, a lot of people watch that," Psaki said in an interview with the left-wing *Pod Save America* in January 2022. "They think that the president isn't doing anything to address people's safety in New York and that couldn't be further from the truth."

Actually, it's the complete truth: When looking at the crime numbers on a macro level, FBI data shows that murders hit a twenty-five-year high in 2021, while sixteen cities hit homicide records last year. At the time Psaki made those comments and even today, polls show about eight in ten voters saying they believe violent crime is a "major problem" in the United States. As for the way the president is seen as doing his job on this issue, Biden polls in the 30s in terms of approval.

That's reality, not an alternate universe.

Arguably the worst tragedy of 2021 occurred in Waukesha, Wisconsin, on November 21. A Christmas parade featuring young children and grandmothers was being held as it is every year in the town located twenty miles outside of Milwaukee, when suddenly a madman mowed down six people and injured sixty-two.

Four of the those killed were members of the Milwaukee Dancing Grannies. The dead ranged in age from eight to eighty-one. The man charged, Darrell Brooks, was out on one thousand dollars' bail after attempting to drive over the mother of his child.

When asked if the president would assume the role of consoler in chief, Psaki offered up this insult.

"Obviously any president going to visit a community requires a lot of assets . . . requires taking their resources," she told reporters on November 29. "And it's not something that I have a trip to preview at this point in time, but we remain in touch with local officials."

Since when has the Biden administration cared about saving resources? The following day, the president would fly over Waukesha to attend a fundraiser in Minnesota while touting his infrastructure plan.

Really. Disgrace doesn't begin to describe it. It's plain that this tragedy didn't merit notice for the Biden administration because there was nothing to exploit politically.

Jen couldn't let reality slip into her worldview. And you know what else is jettisoned from Planet Psaki? Understanding of immigration incentives. You'll *love* this response regarding a question from Fox's Peter Doocy of why migrants crossing over the border aren't being asked about proof of vaccination or a negative COVID test.

"As individuals come across the border, they are both assessed for whether they have any symptoms," Psaki responded. "If they have symptoms, the intention is for them to be quarantined. That is our process.

"They're not intending to stay here for a lengthy period of time," she added.

On what planet are migrants traveling hundreds of miles by foot . . . to enter this country illegally . . . perhaps risking their lives in some cases . . . only to turn around and leave after a short period of time? This ain't a trip to Disney. Even if you somehow buy Psaki's answer—and you'd have to be either insane or insanely drunk to do so—does COVID stay away from those who only stay here for a few weeks? How does that work exactly?

It was also amusing to watch Psaki's press briefings because of the press's upside-down COVID optics. Here we had the whole White House press corps—all vaccinated—masked up every day.

Which is fine given how small the James S. Brady briefing room, once home to an indoor swimming pool, actually is. But for whatever reason that has little to do with science, Psaki, also fully vaccinated, was maskless throughout her time at the podium.

One would think she would be asked about why she's exempt from the rule here, but it never happens. If it's a Republican like Senator Ted Cruz (R-TX), however, then suddenly the press is petrified for its own safety. We witnessed this on March 24, 2021, when Cruz stepped up to a microphone for a press conference inside the Senate without a mask on.

"Would you mind putting on a mask for us?" an off-camera reporter asked.

"Uh, yeah, when I'm talking in front of the TV cameras I'm not going to wear a mask," Cruz responded. "And all of us have been immunized, so . . ."

"It would make us feel better," the reporter pressed.

"You're welcome to step away if you like. The whole point of a vaccine . . . CDC guidance is what we're following," Cruz explained.

It would happen again in January 2022 after another reporter questioned Cruz about him and other Republicans not wearing masks.

"Just once, I'd like to see a reporter say to Joe Biden when he stands at the damn podium in the White House without a mask, 'Mr. President, why aren't you wearing a mask?'" Cruz said. "Just once, I'd like to see you say to Jen Psaki, the White House press secretary, when she stands at the podium with no mask, 'Ms. Psaki, why don't you have a mask?' The questions are only directed at one side, and I got to say that the American people see the hypocrisy."

He was 100 percent correct. But that didn't stop Psaki from injecting herself into public debates about masking. Here's a tweet she directed at Virginia governor Glenn Youngkin on his first day in office after he followed through on a campaign promise to allow parents to decide if they want their child masked or not.

"Hi there. Arlington county parent here," Psaki tweeted at Youngkin. "Thank you to @APSVirginia [Arlington Public Schools] for standing up for our kids, teachers and administrators and their safety in the midst of a transmissible variant."

So if you're keeping score at home, the press secretary didn't want to allow parents the option of masking their kids, and instead believes it's the role of the government to make that decision. Youngkin won as a heavy underdog against Terry McAuliffe largely because of his stance on parental choice on education, on masks. But here was Psaki, an unelected official, making a public show of it, all while doing her job without a mask on. She also ignored the fact that schools are among the safest places to be in the country when it comes to COVID transmission, or lack thereof.

Oh, and if you're wondering why Joe Biden takes so few questions from the press, or why the president will literally admit that he's not supposed to take questions or "he'll get in trouble," look no further than his press secretary at the time.

"He takes questions nearly every day he's out [with] the press," Psaki said while lying again, because we've all seen Biden time and time and time again run away from the press after speaking about X, Y, Z. "A lot of times, we say, 'Don't take questions.' But he's going to do what he wants to do because he's the president of the United States."

Okay, that's fine. Press secretaries try to keep messaging consistent within any administration. If the president contradicts his press secretary or vice versa, that's an unwanted headline. But please stop insulting everyone's intelligence by saying Biden takes questions nearly every day. It's an invalid truth if there ever was one.

This next Psaki interview also serves as a microcosm of the line between the politicians and the press being completely eviscerated through a perpetual revolving door that drives the public insane about the media.

Here we have Psaki, formerly of the Obama-Biden administration, talking to David Axelrod, also formerly of the Obama-Biden administration. Obama leaves office, so both Psaki and Axelrod go to work for CNN, as do multiple other former Obama officials. When Biden takes the White House, Psaki departs CNN to go back in government as the president's press secretary. She proceeds to do an interview with Axelrod, which was really more of a chat between old friends. Psaki then reveals that the president is being told not to speak to the press, and Axelrod doesn't challenge her at all. We saw the same revolving door between media and government under Trump as well and Obama before him. Such is media in the 2020s.

But if you are looking for what was the most embarrassing piece of "journalism" anyone could ever engage in, look no further than Psaki's interview with CNN media critic Brian Stelter.

"What does the press get wrong when covering Biden's agenda, when you watch the news, when you read the news, what do you think we get wrong?" Stelter asked Psaki in June 2021 on a program. Translation: *How can we help serve you better*?

More Stelter: "You've mentioned your kids. You have a daughter going into kindergarten, I have a daughter going into pre-K, and I think to myself what kind of country is this going to be when they are our age? Do you fear that given the craziness we're seeing from the GOP, do you fear that for our kids, your kids and mine?"

Glenn Greenwald, an actual journalist with a Pulitzer Prize to prove it, was not impressed. "I'm not using hyperbole when I say the 'interview' that Brian Stelter did with Jen Psaki yesterday should be studied in journalism school," he wrote to his 1.7 million followers. "It's one of the most sycophantic interviews of a state official you'll ever see. This is how state TV functions."

Joe Rogan, who draws *11 million* listeners per episode, or 22 times the audience CNN averages in a given day, was a bit more direct. "Brian Stelter talking to the press secretary saying 'what are we doing wrong?' What are we doing wrong? Like, hey motherfucker, you're supposed to be a journalist!"

Psaki left the administration in May 2022 for a lucrative hosting job at MSNBC. Talk about a great fit. As for Psaki's issues with the truth, if Rachel Maddow is the one who sets the bar on this front over at MSNBC, getting over it won't be a problem.

It was Maddow, of course, who pushed Trump-Russia collusion more than anyone in the media, which dominated her program for the better part of two years on a nightly basis. Fast-forward to 2022, when Psaki went full-Maddow in declaring, without real evidence, that Russia "hacked" the 2016 U.S. presidential election.

"If you look back at 2014, and frankly even 2016, when Russia invaded Ukraine and then in 2016, when they, you know, of course, hacked our election here, we did not do that. Right? We did not declassify information."

Hacked? That unequivocally implies voting machines were compromised. Votes were changed. None of this happened, of course, according to U.S. intelligence. But if Psaki were to say the same thing on TV, she'd get away with it. She won't be paid to report the news, but rather to spin it in a way that serves up red meat to its core audience of Democrats/liberals/anti-Trumpers.

On her way out the door, Psaki showed how utterly classless and unprofessional she is. During an interview with the podcast *Pod Save America*, she was asked if Peter Doocy was truly a "stupid son of a bitch." The question derived from Biden calling Doocy just that for simply asking a question about inflation and its impact on the midterms. Doocy, for his part, didn't make himself the story or play the victim the way Psaki's former CNN coworker Jim Acosta would have.

"He works for a network that provides people with questions that—nothing personal to any individual, including Peter Doocy—but might make anyone sound like a stupid son of a bitch," she said, drawing laughs from the all-liberal audience.

Let's make one thing abundantly clear: *Never* has anyone told me what to say or what to ask when appearing on Fox News as an analyst or as a host. Anyone who works there now or has worked there in the past would tell you the same exact thing. So again, she's lying. Also, by mocking Doocy as someone who needs his questions written for him, she shows just how unprofessional and condescending she actually is.

For his part, Doocy didn't run to Twitter to hit back at Psaki, didn't whine about it on Fox News. Unlike Acosta, Doocy just asks solid questions and doesn't try to make himself the story. How refreshing.

By the way, without a foil, is Psaki *really* that interesting? She was a run-of-the-mill pundit at CNN from 2017 to 2020. Her clips and analysis rarely made headlines.

In her new role, she'll need to provide one hour of compelling content *five days a week*. There won't be a Peter Doocy or Jacqui Heinrich, both Fox News White House correspondents who stand out as two reporters who actually challenged Psaki with tough questions, on set there with her to make matters more interesting.

If Psaki does what Maddow does—engage in long monologues followed by like-minded guests—it's difficult to see that drawing much better numbers than the limp results the networks are drawing now. Her presentation is awkward. She doesn't come across as authentic. She's profoundly condescending.

But she's a legitimate star among left-leaning media members and those who follow them on social media. And given the lack of quality choices for MSNBC, it made sense to offer her a contract.

Psaki's replacement was Karine Jean-Pierre, formerly of MSNBC. Jean-Pierre could be considered the most unsteady, unprepared person to ever serve as White House press secretary in the history of the position. She reads many of her answers to questions directly off notecards, sometimes for paragraphs at a time. And those answers occasionally aren't related to the question asked.

The best example of this came in May when Doocy quoted a Biden statement arguing that the way to lower inflation is to raise taxes on corporations.

"How does raising taxes on corporations lower the cost of gas, the cost of a used car, the cost of food, for everyday Americans?"

What followed sounded like an unprepared sophomore in high school desperately seeking to achieve a mandatory word count on a book report. Here was the answer verbatim, per the official White House transcript:

Jean-Pierre: So, look, I think we encourage those who have done very well—right?—especially those who care about climate change, to support a fairer tax—tax code that doesn't change—that doesn't charge manufacturers' workers, cops, builders a higher percentage of their earnings; that the most fortunate people in our nation—and not let the—that stand in the way of reducing energy costs and fighting this existential problem, if you think about that as an example, and to support basic collective bargaining rights as well. Right? That's also important. But look, it is—you know, by not—if—without having a fairer tax

> code, which is what I'm talking about, then all—every—like manufacturing workers, cops—you know, it's not fair for them to have to pay higher taxes than the folks that—who are—who are—who are not paying taxes at all or barely have.

Feel free to read that five times. It makes less and less sense with each attempt. But Jean-Pierre checked off several boxes when it came to race and sexual orientation, so she got the job over an infinitely more worthy John Kirby, who served as Pentagon spokesperson at the time.

One person who probably wishes she was joining Psaki at the Peacock is Kamala Harris. Because the way things are going for the vice president, any other job with a celebrity aspect would likely be welcome right about now. And not just to Kamala, but the island of misfit cabinet members about to be installed into positions they had no business being chosen for.

8

WE WERE PROMISED ADULTS IN CHARGE

"The adults are back in charge." That was the narrative we heard entering the Biden era. The qualifications for a job as an Biden administration official seemed to be this: be diverse and be an adult.

"Fulfilling a Promise: A Cabinet That Looks Like America," swooned the *New York Times*. "Joe Biden's Cabinet Picks Send a Clear Message: The Adults Are Back in Charge," declared *Vogue*. "Joe Biden's message to Vladimir Putin? The adults are back in charge," CNN editor at large Chris Cillizza echoed after Biden first met with Vladimir Putin.

"While Biden didn't seek to fully explain what it was he came to do, the evidence was everywhere in his answers: Make clear that the traveling circus of the Trump presidency was over—and that the adults were back in charge," Cillizza wrote for CNN.com. "Nuts and bolts. Pragmatism over pomp. Total frankness over Trumpism. Or, in Biden's words: 'This is about how we move from here. This is—I listened to, again, a significant portion of what President Putin's press conference was—and as he pointed out, this is about practical, straightforward, no-nonsense decisions that we have to make or not make.'"

Quite the press release, er, analysis. In a related story, here's Cillizza in 2016: "Let me say for the billionth time: Reporters don't root for a side. Period."

So the country could breathe easier again because Team Trump was out and the experienced professionals were taking the wheel. *Order* would be restored, and normalcy would rule once again.

But who were these adults, exactly? Had any serious journalists or publications vetted Biden's choices to run Transportation, Energy, Homeland Security, or State? It turns out that Biden's cabinet choices were . . . over eighteen, but that didn't equate to being adults.

Let's start with the Department of Transportation and Pete Buttigieg, the former mayor of the college town of South Bend, Indiana. During the 2020 primaries, after winning Iowa and pummeling Biden in New Hampshire in *tripling* his vote total, Mayor Pete was kind enough to get out of Joe's way before Super Tuesday by dropping out of the race and endorsing him.

But instead of Biden giving Buttigieg some kind of symbolic job, like ambassador to New Zealand, he was handed the vast Department of Transportation. This is the same guy who had trouble filling potholes in South Bend.

From *Politico Magazine*, December 2020:

> About 22,000 [potholes] gape across the city's streets each year. In January of 2019, residents of the city asked Domino's Pizza for help to augment the city's $90,000 pothole budget and won a $5,000 grant for its public works department, the only town in the state of Indiana to do so.

You read that right. *Domino's Pizza* had to intervene here on behalf of South Bend residents to help address an issue Mayor Pete was having difficulty with. Overall, South Bend is a city with just eighteen bus routes and sixty buses. It has one train station and one airport that would never be confused with O'Hare. *So who better* to run Transportation, which consists of 58,000 employees and a budget of $87 billion, than Mayor Pothole himself?

In the summer of 2021, a supply chain crisis exploded across the country that impacted (and still impacts) many Americans, particularly the lower and middle classes. Restaurants, stores, and small businesses that provide products and services to people were all negatively impacted as the cost to the consumer continued to rise. And when the going got tough that summer, the tough went on paternity leave. Which is fine, of course. Being a father of two myself, it is totally understandable to want to spend time with those little new members of the family.

Buttigieg, also a media darling in what appears to be a pattern among many Democrats and the press, got quite the treatment on becoming a father of two along with his husband, Chasten.

> *People*: "Pete Buttigieg Calls Parenting Twins 'Most Demanding Thing': 'Yet I Catch Myself Grinning Half the Time'"
>
> *USA Today*, "Celebrities" section: "Pete Buttigieg calls parenting twins 'the most demanding thing I think I've ever done'"
>
> NBC News: "Buttigieg on parenthood: 'Most demanding thing' I've ever done"

But this wasn't your ordinary paternity leave. Because when leaving a major cabinet post for something as important and crucial as the Department of Transportation, especially during a supply chain crisis, one would think that 1) an acting secretary would be named, and 2) Buttigieg might want to let the American public know that he'd be out for (checks notes) *two months*.

Politico's editor in chief, Matthew Kaminski, tweeted a quote from a *Politico* article by Alex Thompson and Tina Sfondeles on October 14, 2021, "They didn't previously announce it, but Buttigieg's office told West Wing Playbook that the secretary has actually been on paid leave since mid-August to spend time with his husband, Chasten, and their two newborn babies."

They didn't previously announce it? Amateur hour, for sure.

Buttigieg, who is barely forty, will have to defend himself on the supply chain front (which isn't going away anytime soon) and about his work ethic in general if he decides to run for president in 2024 if his boss decides to bow out. He therefore cannot be taken seriously as a candidate anytime soon.

Okay, let's go over to the Department of Energy, another hugely important cabinet position. As noted earlier, Biden chose Jennifer Granholm, former Michigan governor and CNN political analyst. One afternoon in November 2021, when gas prices were surging, Granholm responded to an interview question by laughing after being asked what she would do to address those rising prices.

"What is the Granholm plan to increase oil production in America?" Bloomberg's Tom Keene asked.

"That is hilarious!" Granholm cackled in response before adding something in semi-English. "Would that I had the magic wand on this!"

Keene, to his credit, didn't laugh along. Granholm suddenly got nervous in her realization that she might actually have to provide a real answer.

"As you know, of course, oil is a global market," she argued. "It is controlled by a cartel. That cartel is called OPEC, and they made a decision yesterday that they were not going to increase beyond what they were already planning."

Hmmm . . . so does OPEC control U.S. energy production here

at home? And if OPEC is so powerful, why was gas in the $2.00 per gallon range during Trump's presidency?

The Granholm clip went viral, as it should have. So the powers that be from the White House communications team gave her essentially a do-over in joining the daily White House press briefing a few days later. Somehow, she performed *worse* after (again) not being able to answer a simple question:

Reporter: How many barrels of oil does the U.S. consume each day?
Granholm: I don't have that number in front of me.

So, if you're keeping score at home, *the energy secretary of the United States* didn't know how many barrels of oil the U.S. consumes on a daily basis (answer: 18 million). One must wonder what experience Granholm had to get this important job in the first place.

Then there's Secretary of Homeland Security Alejandro Mayorkas, the guy who still insists up and down that the border is closed while he somehow simultaneously blames the Trump administration for a surge in migrants that will result in more than 2.3 million illegally entering the country in 2022. Add that to the more than 2.3 million who entered in Biden's first year, and that's more than 4 million, or more than *six times* the population of Boston.

"Former president Trump [. . .] slashed the resources that we were contributing to address the root causes of irregular migration," Mayorkas claimed in August 2021 as illegal border crossings were completely out of control. That month, according to U.S. Customs and Border Protection statistics, there were more than 200,000 migrant encounters with Border Patrol. Ultimately, that year would end with a total of 2,035,595 migrant encounters, topping even the previous record from 2000, according to Pew Research. Throw in 300,000 "gotaways" (those not encountered and got into the country) and we're talking more than 2.3 million people entering this country illegally through the U.S. southern border alone.

"The policies have changed so dramatically between the Trump administration and the Biden administration. We have ended policies of cruelty that defined the prior administration," he also claimed in January 2022. Again, this is the person tasked with safeguarding the American people.

So blaming Trump for the border catastrophe takes serious brass. It's also a messaging strategy that smells as bad as it sounds. According to the RealClearPolitics average of major polls, Biden-Harris-Mayorkas is polling at just 27 percent approval on immigration, or 33 points underwater. When limiting the question to handling of the border, Biden-Harris-Mayorkas clocks in at 23 percent. Boy, I'd love to play cornhole with that 23 percent over a few suds sometime . . .

Fortunately, border officials and agents have had enough. One January 2022 meeting got tense in a hurry.

"You said you're going to change it. The only thing that's changed is it's gotten worse," one agent in Texas told Mayorkas, according to audio obtained by TownHall.com. At another meeting, an agent turned his back on the homeland head.

"You can turn your back on me, but I'll never turn my back on you," Mayorkas reportedly said in response.

"[I] heard Border Patrol agents loud and clear. . . . I am fighting to get more resources they need," Mayorkas also tweeted in January 2022.

Actually, that's not the case. In a federal budget unveiled in April 2022, Biden allocated $309 million for border security technology and $494 million for "noncitizen processing and care costs." So we're talking just $800 million out of $5.6 trillion, *0.014 per-*

cent of the entire budget. Nothing for border wall construction, of course.

In the announcement of Mayorkas's nomination to run homeland security, what many in the press focused on wasn't Mayorkas's qualifications to lead a 240,000-person department, but (drumroll) *diversity*! And the fact that he supposedly cares more than those sinister Trump border folks (who kept said border under control).

The *Washington Post* wrote this before Mayorkas's confirmation:

> Mayorkas, 61, is a former federal prosecutor, not a liberal activist, but he brings a deep sympathy for immigrants rooted in his own family's extraordinary journey to the United States, his backers say. He would be the first Latino, and first immigrant, in charge of DHS, the sprawling domestic security agency created after the 9/11 attacks to combat terrorism, safeguard ports and borders, and enforce immigration laws inside the United States.

It's the check-boxes cabinet, clearly.

Then, of course, there's also Secretary of State Antony Blinken. Before you say, "Hey, how could we have known he would be this bad" . . . simply soak in these quotes from the late senator John McCain in 2014 when he opposed Blinken's nomination for deputy secretary of state.

"This individual has actually been dangerous to America and to the young men and women who are fighting [for] and serving it," McCain said of Blinken, referring to his foreign policy worldview as "at worst anti-strategic."

McCain warned that setting concrete timetables for an Afghanistan withdrawal was foolish, and he advocated leaving behind a "stabilizing force" in Afghanistan. Blinken opposed both.

McCain added, "Mr. Blinken said, 'We've been very clear. We've been consistent. The war [in Afghanistan] will be concluded by the end of 2014. We have a timetable and that timetable will not change.' This is why I'm so worried about him being in the position that he's in. Because if they stick to that timetable, I'm telling my colleagues that we will see the replay of Iraq all over again. We must leave a stabilizing force behind of a few thousand troops."

Fast-forward to 2021 and somehow Tony Blinken is in charge of the State Department. Just as McCain predicted, Blinken and Biden pulled out all U.S. troops from Afghanistan in August 2021 by sticking to a firm timetable and broadcasting that commitment to the world. We all remember the horrifying images of mobs pressed up against barricades, with desperate people throwing children over the barriers. The results were fatal for U.S. service members, with thirteen being killed at the airport. Americans were left behind. A ragtag Taliban called the shots during the withdrawal. Overall, it was an utter embarrassment on the world stage that only emboldened the Putins and Xis of the world.

We should have seen this coming during Blinken's confirmation hearings in early 2021. Because *this* is what he said would be the epicenter of U.S. foreign policy and national security.

"We'll put the climate crisis at the center of our foreign policy and national security, as President Biden instructed us to do in his first week in office," he said. He was serious. "That means taking into account how every bilateral and multilateral engagement—every policy decision—will impact our goal of putting the world on a safer, more sustainable path."

This bears repeating: "We'll put the climate crisis at the center of our foreign policy and national security." No wonder the press barely put a glove on this guy before he was confirmed: He was the wokest secretary of state this country had ever seen.

Under Blinken's watch, McCain proved to have been 100 percent correct. Antony Blinken is dangerous to this country and our allies. The Taliban runs Afghanistan. ISIS and Al Qaeda have a playground to plan attacks on our homeland again. North Korea has resumed earnest missile tests. Russia invaded Ukraine without provocation. China is eyeing up Taiwan. And what is Blinken doing? Trying to revive the Iran nuclear deal.

We're governed by children. By a political appointee who should be paternity czar instead of transportation secretary. By an energy secretary who should be back on CNN instead. By a homeland security chief who should be working in an E-ZPass booth. By a secretary of state who should have George Costanza's job as assistant to the traveling secretary instead.

Sleep well, readers!

9

MILLI BIDENILLI

The year was 1987. How far back is '87 looking these days? A few fun facts: The Fairness Doctrine, which paved the way for something called *The Rush Limbaugh Show* to debut on radio one year later, had just been revoked.

Moonlighting, starring the great Bruce Willis (who had hair then) and Cybill Shepherd, was one of TV's hottest shows, on ABC. How hot? One episode in March 1987 drew *60 million viewers*. For context, Biden's inaugural address in 2021 drew just over half that number (33.8 million) despite being carried by seventeen networks.

In 1987, inflation was at 3.65 percent. In 2021, it averaged 7 percent. A gallon of gas was 95 cents (it's just a *bit* higher today). The federal minimum wage was $3.65. The Minnesota Twins, who once played indoors in a stadium nicknamed the Homerdome, won the World Series for the first time since 1924. "Walk Like an Egyptian" by the Bangles, with the lovely Susanna Hoffs on lead vocals (by far my teenage crush to end all crushes), was the number one song of the year.

You get the idea. It's hard to believe, but 1987 was somehow already thirty-five years ago. *Everything* seems as ancient as an answering machine and New Coke. And it was that year that a not-so-ancient senator named Joe Biden declared he was running for the highest office in the land at just forty-four years old. If he had won in 1988, Biden would have been the youngest president in

U.S. history (ironic considering he eventually became the *oldest* president in U.S. history).

In making his announcement, the 1987 version of Biden provided the following perspective. It's a doozy when considering where we are now.

"There are risks we must take in foreign policy and national security if we are going to shape our children's world. America cannot retreat from the world," 1987 Biden tried to warn 2021 Biden about Afghanistan. "We cannot succumb to the isolationist instincts of those who would put up trade walls to keep out the world, or others who would pull a Star Wars cover over our heads—a modern 'Maginot Line'—ravaging our economic capital, nuclearizing the heavens, and yielding the fate of our children's world to the malfunction of a computer. Like it or not, our only choice is to compete and prosper in the world beyond our shores."

Yup. Walls were *bad* then. As was Star Wars, which was a missile defense system the Soviets believed could protect the U.S. during a nuclear war. President Reagan and his administration used Star Wars masterfully from a messaging and sales perspective, even rigging successful tests to scare the shit out of the Kremlin. The Soviets went on to spend so much to beef up their military capabilities because of it, it helped bankrupt the country, leading to its downfall.

But overall, given the competition for the 1988 Democratic nomination that included heavyweights such as Massachusetts governor and future Tank Boss Michael Dukakis, civil rights leader Jesse Jackson, Senator Al Gore of Tennessee, and Senator Paul Simon of Illinois, who somehow didn't win *People*'s Sexiest Man Alive award that year, Biden had as good a chance as any to take on George H. W. Bush, Ronald Reagan's vice president, for the presidency.

"I looked around, judged myself against the other potential

candidates for the nomination, and by the beginning of 1987 I decided I could beat them," Biden said at the time.

Then, less than four months after announcing his presidential run, Biden was out. But not because of a lack of support or a health issue, but due to a plagiarism scandal that, in a sane world, should have made this bid for the White House his last. Typical of the Biden we see today, he didn't show remorse or regret despite overwhelmingly clear evidence that he had lifted parts of speeches from a British politician named Neil Kinnock, the British Labor Party leader who ran unsuccessfully against Prime Minister Margaret Thatcher. Instead, Biden portrayed himself as a helpless target of dirty politics.

"Although it's awfully clear to me what choice I have to make, I have to tell you honestly, I do it with incredible reluctance, and it makes me angry," Biden grumbled during an announcement that officially ended his run in September 1987.

"I'm angry at myself for having been put in the position—put myself in the position—of having to make this choice," he continued. "And I am no less frustrated for the environment of presidential politics that makes it so difficult to let the American people measure the whole Joe Biden and not just misstatements I have made."

Yep. Biden was a victim of the environment of presidential politics, dontcha see? He actually admitted to the plagiarism while saying it was not "malevolent."

Judge for yourself. It's the Milli Vanilli of plagiarism.

> **Neil Kinnock, May 1987:** Why am I the first Kinnock in a thousand generations to be able to get to university? Why is Glenys [his wife] the first woman in her family in a thousand

generations to be able to get to university? Was it because all our predecessors were thick?

Joe Biden, August 1987: Why is it that Joe Biden is the first in his family ever to go to a university? Why is it that my wife [Jill Biden] who is sitting out there in the audience is the first in her family to ever go to college? Is it because our fathers and mothers were not bright? Is it because I'm the first Biden in a thousand generations to get a college and a graduate degree that I was smarter than the rest?

Kinnock in 1987 on Welsh coal miners: Did they lack talent? Those people who could sing and play and recite and write poetry? Those people who could make wonderful beautiful things with their hands? Those people who could dream dreams, see visions? Why didn't they get it? Was it because they were weak? Those people who could work eight hours underground and then come up and play football? Weak?

Biden in 1987 on ancestors who he claimed worked in the coal mines of northeast Pennsylvania: Those same people who read poetry and wrote poetry and taught me how to sing verse? Is it because they didn't work hard? My ancestors, who worked in the coal mines of northeast Pennsylvania and would come up after twelve hours and play football for four hours?

Well, the good news for Biden is that the football reference was seamless. Sure, Kinnock was talking about soccer, but American football is a big effing deal in Pennsylvania. Joe Namath, Joe Montana, Dan Marino . . . Skip "Bulldog" Biden.

So this isn't just plagiarism with a small *p*. It's rhetorical grand theft auto. It's also an outright lie that Biden's relatives worked in coal mines. They didn't. So, yes, somehow Biden combined Kinnock's fact-based family background and combined it with his fabricated one. It's almost impressive on some level.

If you are looking for a good laugh: When Biden adviser Mike Donilon was asked at the time about which of Biden's ancestors were coal miners, he went the full Psaki: "[They were] people that his ancestors *grew up with* in the Scranton region, and in general the people of that region were coal miners."

Ah. Bloodline by association.

> **Kinnock, May 1987**: Does anybody really think that they didn't get what we had because they didn't have the talent or the strength or the endurance or the commitment? Of course not. It was because there was no platform upon which they could stand.

> **Biden, August 1987**: No, it's not because they weren't as smart. It's not because they didn't work as hard. It's because they didn't have a platform upon which to stand.

At the end of the speech, Biden didn't attribute any of it to Kinnock. Because he wasn't being malevolent or anything . . .

Fortunately, 1987 was a time when late-night comics actually didn't play the role of Democratic activists. The King, Johnny Carson of NBC's *Tonight Show*, who was extremely selective over when to make a political joke and never favored a side, had a hammer waiting for Biden.

"On the political scene, one of the Democratic candidates, a Senator Joseph Biden—have you seen the problem he's been having? He went around and made a speech and apparently he quoted a—I think it was a British politician, took his speech and kind of paraphrased it as his own.

"And then the press got on him. And then he was charged also with taking part of Bobby Kennedy's speeches. And Biden says, not to worry. He reassured his staff, he said, 'We have nothing to fear but fear itself.'"

Couldn't you, like, *totally* hear Stephen Colbert or Jimmy Kimmel roasting Biden like that? And this wasn't the only time: Biden's first recorded plagiarism incident came twenty-two years earlier at Syracuse Law School. Again, he swore it was just an accident. And this wasn't just lifting a few paragraphs, either, but *five pages* used in a fifteen-page report.

"I was stupid. I did not intend to mislead anybody. I thought what I was doing honestly was the right way to do it," Biden explained in 1987 after the Syracuse incident came to light.

Biden must be Irish, because instead of being thrown out of law school, Biden was lucky enough to be given the chance to take the class over again. Go Orange! So as you can see, we're dealing with a chronic plagiarist and perpetual liar going back at least *fifty-seven years*. But he wouldn't dare go down this road in the twenty-first century, after all these actions have cost him in the past, right? *Right?*

Well, let's go to the year 2008. Senator Biden is now well into his sixties at this point. Older. But not wiser. Here's Biden speaking on a resolution honoring South Korean president Lee Myung-bak, which would be inserted into the Congressional Record:

Biden, February 2008: He told the city's people that he would remove the elevated highway that ran through the heart of Seoul and restore the buried Cheonggyecheon stream—an urban waterway that Lee himself had helped pave over in the 1960s. His opponents insisted that the plan would cause traffic chaos and cost billions. Three years later, Cheonggyecheon was reborn, changing the face of Seoul. Lee also revamped the city's transportation system, adding clean rapid-transit buses.

***Time* magazine, Bryan Walsh, October 2007:** He told the city's people that he would tear out the jam-packed elevated highway that ran through the heart of Seoul and restore the buried Cheonggyecheon stream—a foul urban waterway that Lee himself had helped pave over in the 1960s. His opponents insisted that the plan would cause traffic chaos and cost billions, but the voters elected Lee. Three years later, Cheonggyecheon was reborn, an environmentally friendly civic jewel that has changed the face of Seoul. More quietly, Lee also revamped the city's transportation system, adding clean rapid-transit buses.

Penalty? Consequences? Nope. Barack Obama would choose him as his running mate a few months later. Think about this for a moment: Joe Biden has been a guy who usually padded his résumé, boosted his hardscrabble credibility, and portrayed himself as a knight in shining armor. This was *always* a lie. And the Democrats always forgave him for it.

As a presidential candidate, with his advisers clearly not giving two shits if they got caught lifting other people's work because the press would largely give him a pass anyway, Biden introduced his not-so-signature "Build Back Better" slogan in July 2020. In a related story, the United Nations had introduced its climate change initiative, Build Back Better, in April 2020.

This is who Joe Biden is. This is the very Swamp that Donald Trump constantly hammered home about. Yet, because much of our media is broken and now filled with activists posing as journalists, Biden not only got away with all of this to eventually occupy the highest office in the land, he did it while being portrayed as a truth teller.

"We choose truth over facts!" Biden once explained, which may explain his allergy to facts.

The hubris on this guy is something to behold. A close second belongs to his former press secretary.

"If the president were standing here with me today, he would say he works for the American people," Jen Psaki said on Day 1 of the Biden presidency. "I work for him. So I also work for the American people, but his objective and his commitment is to bring transparency and truth back to government, to share the truth, even when it's hard to hear. And that's something that I hope to deliver on in this role as well."

Know this: Joe Biden will say or do anything that is politically expedient. If you think playing the race card is beneath this crusader for transparency and truth, well, you haven't been paying attention.

10

SILVER SPOON BIDEN FAILS UPWARD

His name was Frank Fahey and he lived in Claremont, New Hampshire. Fahey, a teacher, had just listened to candidate Joseph R. Biden ramble for forty minutes, at times with sentences without periods or commas, about why he should be the next president of the United States.

When Biden was finally finished, Fahey asked a perfectly reasonable question in a respectful manner. "What law school did you attend and where did you place in that class?"

This oddly set something off in the senator from Delaware, who had been in this position of power for the past fifteen years and therefore wasn't exactly a novice speaking with voters on the campaign trail.

"I think I have a much higher IQ than you, I suspect," Biden responded while pointing his finger at Fahey. Then he launched into this bizarrely petty rant:

> I went to law school on a full academic scholarship, the only one in my class to have full academic scholarship. The first year in law school, I decided I didn't want to be in law school and ended up in the bottom two-thirds of my class. And then decided I wanted to stay and went back to law school and, in fact, ended up in the top half of my class. I won the international moot court competition. I was the outstanding

> student in the political science department at the end of my year. I graduated with three degrees from undergraduate school and 165 credits; you only needed 123 credits. And I'd be delighted to sit down and compare my IQ to yours if you'd like, Frank.

What kind of adult *speaks like this*? It's like listening to an over-served senior in high school outside of the Peach Pit after midnight begging the cheerleading captain to go to the prom with him.

It was also complete . . . what's the word we're looking for here? *Malarkey.*

Let's break it down claim by claim:

Claim: "I went to law school on a full academic scholarship."
Fact: Biden went to school on a half-scholarship based solely on financial need. Because when you are earning all Cs and Ds in your first three semesters at the University of Delaware and an F in ROTC, full scholarships tend to not be extended.
Claim: "I ended up in the top half of my class."
Fact: Biden finished 76th out of 85.
Claim: "I won the international moot court competition."
Fact: He didn't.
Claim: "I graduated with three degrees from undergraduate school."
Fact: Biden graduated with one degree.

Let's look at how exactly a seventy-something Biden was even in position to run for president in 2020 given his track record on everything from policy to performance to trustworthiness. For instance, in 2007 Joe Biden decided to try again. He was itching to get back into another presidential primary twenty years after his plagiarism debacle had sidelined him. Voters, particularly the press, would forgive him, he figured. He had served in the Senate for thirty-four years at this point. In his mind, he had earned the right to be considered a serious presidential candidate.

"I'm running for president because I think that, with a lot of help, I can stem the tide of this slide and restore America's leadership in the world and change our priorities," Biden told reporters in a conference call in January 2007. "I will argue that my experience and my track record—both on the foreign and domestic side—put me in a position to be able to do that."

It's hard to figure out what track record, exactly, Biden was talking about.

Senator Biden, however, was opposed to that idea.

"I would respectfully suggest to you that the Democrats out there understand I am the only person with a plan that can get out of Iraq without our interests in the region not falling apart," Biden told reporters at the time.

Fourteen years later, Biden's plan as president to get us out of Afghanistan without our interests in the region falling apart failed spectacularly. Biden's 2008 campaign would never get off the ground. On January 3, 2008, after capturing less than 1 percent of the vote in Iowa for a fifth-place finish, it was over as far as Biden's chances to ever get to the West Wing were concerned. Or was it?

Hillary Clinton's coronation as the party's nominee was not going well, all thanks to a relative unknown named Barack Obama. The Illinois freshman senator had been elected to his first term in 2005, but the guy *really* knew how to give a speech and made Hillary look too establishment, too entitled, too phony, too old. Obama would shock the country in winning Iowa, with Hillary finishing third behind somebody named John Edwards (remember him?). The two would trade victories back and forth, with Hillary taking New Hampshire and Obama winning South Carolina. But Obama began to outraise and outhustle Hillary as the race wore on. And more important, he was winning the hearts and minds of the totally objective national press.

"Media Narrative Vaults Obama into Frontrunner Slot," read the headline from Pew Research, which did its usual excellent work in studying coverage of presidential races.

Here are a few examples of the tone of coverage Pew used to underscore its analysis. It would provide a preview of what the great journalist and media observer Bernard Goldberg wrote about in the bestselling *A Slobbering Love Affair: The True (and Pathetic) Story of the Torrid Romance Between Barack Obama and the Mainstream Media.*

Pew highlighted these excerpts from press coverage in a February 19, 2008, article:

> CBS News: "With soaring rhetoric, Obama is moving his audiences not just politically, but emotionally. The stoic eloquence channels John F. Kennedy."
> Associated Press: "Barack Obama could hardly have had a better weekend," said the story, referring to his weekend wins ahead of the Potomac Primary. "For a cherry on top, he won a Grammy award Sunday, beating former president Bill Clinton and others for 'best spoken word album,' for the audio version of his book, 'The Audacity of Hope.'"
> Associated Press: "Memo to Hillary Rodham Clinton: Barack Obama is stealing your faithful."

You get the idea. But Obama's experience was seen as an issue. Being a community organizer didn't exactly inspire confidence when dealing with national security issues in the wake of 9/11 or confronting America's adversaries overseas at a time when the country was fighting two wars and facing a growing threat from North Korea.

After Obama captured the nomination to face war hero John McCain, a critical choice needed to be made regarding who his running mate would be. In his aforementioned book, *The Audacity of Hope*, Obama said the choice came down to Senator Tim Kaine of Virginia (Hillary's future vice presidential choice) and Senator Joe Biden of Delaware.

On August 23, 2008, Obama had made his selection.

"You are the pick of my heart, but Joe is the pick of my head," Obama told Kaine, according to people with knowledge of the exchange, according to the *New York Times*.

The press, as predictable as the Knicks not competing for a championship, ate it all up in portraying Biden as the greatest foreign policy mind since Henry Kissinger, despite his being on the wrong side of the Gulf War in 1991 (voted against) and the Iraq War in 2003 (voted for).

The *New York Times* report on August 23, 2008, read, "In Mr. Biden, Mr. Obama selected a six-term senator from Delaware best known for his expertise on foreign affairs (Mr. Biden spent last weekend in Georgia as that nation engaged in a tense confrontation with Russia) but also for his skills at political combat."

Are they talking about the same Joe Biden?

"Barack Obama's choice of Delaware Senator Joe Biden as the Democratic vice presidential nominee instantly bolsters the ticket's credentials on foreign policy, an area where Obama's background is limited. Biden, chairman of the Senate Foreign Relations Committee, is one of Congress' most knowledgeable and respected voices on national security," wrote ABC News glowingly, while quoting exactly zero lawmakers to back up the whole *most-knowledgable-and-respected-voices-on-national-security* thing.

Something else not emphasized very much in the reporting were statements Biden had made about Obama while he was still a candidate.

"I mean, you got the first mainstream African American who is articulate and bright and clean and a nice-looking guy," Biden had said. "I mean, that's a storybook, man!"

"Mainstream African American."

"Articulate."

"Bright."

"Clean."

Picture *anyone* remotely right of center saying that of any Black guy. The pitchfork protest store would run out of pitchforks to protest with.

But at least Biden was comfortable with Obama as a commander in chief, right?

"I think he can be ready [to be president] but right now I don't believe he is," Biden said during a 2007 primary debate. "The presidency is not something that lends itself to on-the-job training."

As for Biden's thoughts on John McCain, here's what he said in 2005: "I would be honored to run with or against John McCain because I think the country would be better off."

Biden would get the "against" part. Sort of. As a vice presidential candidate.

Obama-Biden would go on to win, of course. But it's hard to imagine how the Democratic Party could have possibly effed up being victorious in 2008, as McCain, to Republicans at the time, was the flat-beer candidate. It's like the old days of tailgating in college after a game. Whatever beer that was brought to the occasion is seven hours old in the postgame. But with no other options, you accept the flat beer because there's no other good choice. Hence the flat-beer candidate of 2008. (Are we out of Ronald Reagan? Fine, guess I'll have a John McCain.)

No election had lined up more favorably for one party:

- Bush 43, the sitting president, was at *25 percent approval* going into Election Day. (Richard Nixon was at 24 percent approval the day he left office after being forced to resign.)
- The 2008 crash occurred two months before Election Day. The economy was in shambles. The party in power in the White House always gets the blame.
- McCain was a decent presidential candidate but could not get any real traction with the media. Journalists were either openly cheering for Obama or obsessed with McCain's choice for running mate, Alaska governor Sarah Palin (and not in a good way).

So basically, Obama could have chosen Joe the Plumber or Joe Lieberman or Joe Torre as his running mate, and the ticket was still going to win easily. But no matter: Joe Biden finally got to the White House.

Next up, the media established a fantasy that Obama and Biden weren't just political colleagues, but like, besties, 4real. *Politico Magazine* ran a piece in August 2020 during the Democratic National Convention detailing the Obama-Biden relationship. On the outside, they went out of their way to show *just how tight* buds they were. Because every guy I know, whether it's on my softball team or in my fantasy football league, buys a *best friend bracelet* for *their* BFF, complete with smiley faces and flowers. Totally normal.

"Happy #BestFriendsDay to my friend, @BarackObama," Biden's PR team wrote on Twitter in June 2019.

But according to *Politico*, things were festering internally, as Biden's continued gaffes were mocked by younger Obama staffers.

"You could certainly see technocratic eye-rolling at times," said Jen Psaki, who was White House communications director at the time. Psaki, of course, would go on to be Biden's first press secretary.

Obama at times felt a need to placate his vice president in meetings, only to ultimately ignore him, according to then–FBI director James Comey in his memoir, *A Higher Loyalty*.

"Obama would have a series of exchanges heading a conversation very clearly and crisply in Direction A. Then, at some point, Biden would jump in with, 'Can I ask something, Mr. President?'" Comey recalled.

"Obama would politely agree, but something in his expression suggested he knew full well that for the next five or 10 minutes we would all be heading in Direction Z. After listening and patiently waiting, President Obama would then bring the conversation back on course."

"In the Situation Room, Biden could be something of an unguided missile," Obama deputy national security advisor Ben Rhodes wrote in his memoir.

Overall, Biden was at odds with his boss and top military brass on major issues: Obama and his defense secretary, Robert Gates, were for a troop surge in Afghanistan, but Biden was against it. Obama and Gates were both for military intervention against Libya and Muammar Gaddafi, but Biden was against it. On the raid that took out Al Qaeda leader Osama bin Laden, Obama and Gates were for it, while Biden said we shouldn't go in.

"I think he has been wrong on nearly every major foreign policy and national security issue over the past four decades," Gates said of Biden in his book *Duty: Memoirs of a Secretary at War* in 2014, while Biden was still vice president.

Biden had also—on several occasions—publicly embarrassed his BFF on domestic issues. On gay marriage, the devout Catholic came out for it without warning in an interview on NBC's *Meet the Press* while his boss was still publicly opposed.

"I—I—look, I am vice president of the United States of America," he told then-moderator David Gregory in 2012. "The president sets the policy. I am absolutely comfortable with the fact that men marrying men, women marrying women, and heterosexual men and women marrying another are entitled to the same exact rights, all the civil rights, all the civil liberties. And quite frankly, I don't see much of a distinction beyond that."

According to the book *Game Change*, by John Heilemann and Mark Halperin, the reaction inside the White House was about what one would expect.

"What the fuck? How can this have happened?" Obama's senior adviser David Plouffe responded upon reading the transcript.

Biden also stepped all over Obama's March 2010 ceremony for his signature achievement, the Affordable Care Act, commonly known as Obamacare. After introducing the president, Biden embraced Obama with a hug before whispering in his ear as loudly as Tom Brady calls out signals before a snap.

"This is a big fucking deal," Biden said, which could very clearly be picked up on a nearby microphone.

Obama, perturbed, responded by looking down and patting Biden twice on the shoulder, almost to usher him off the stage, as if to say, "Please don't fuck this one up for me, too." The ceremony, consequently, became as much about Biden as it was Obama in news reports that day.

As Obama began his reelection effort in 2012, serious discussions were being held about replacing Biden with Hillary as his running mate.

"President Obama's top aides secretly considered replacing Vice President Joseph R. Biden Jr. with Hillary Rodham Clinton on the 2012 ticket, undertaking extensive focus-group sessions and polling in late 2011 when Mr. Obama's re-election outlook appeared uncertain," the *New York Times* reported in a 2013, after-the-election story. "The aides concluded that despite Mrs. Clinton's popularity, the move would not offer a significant enough political boost to Mr. Obama to justify such a radical move, according to a newly published account of the 2012 race."

Halperin and Heilemann's bestselling book on the 2012 presidential election, *Double Down*, underscored just how serious the replace-Biden-with-Hillary effort was.

"When the research came back near the end of the year, it suggested that adding Clinton to the ticket wouldn't materially improve Obama's odds," they wrote. "Biden had dodged a bullet he never saw coming—and never would know anything about, if the Obamas could keep a secret."

As we all know, DC is the opposite of that old Vegas mantra about things happening there staying there. When it happens in the Swamp, it *never* stays in the Swamp. The secret about replacing Biden got out. Of course, that was all by design.

By the way, in retrospect, Obama wasn't exactly in a desperate position to consider such a radical move of replacing his vice president: Osama bin Laden was killed in 2011. The economy was plodding along far too slowly but wasn't horrible (Obama would use the recession he inherited as a crutch during the entire 2012 campaign, as if he were running against Bush more than Romney).

Overall, things were relatively drama-free (when compared to 2016 and 2020) in the year of the 2012 election. Obama's approval rating heading into Election Day was at 52 percent after starting the year in the mid to high 40s, and Mitt Romney wasn't exactly Mr. Excitement on the campaign trail. The fact that the Obama campaign had *even considered* replacing Biden as seriously as it did spoke volumes about how expendable his vice president was considered internally.

Still, heading into 2016, one would think it would only be natural that Obama's vice president would succeed him. That's what usually happens in these situations, save for Dick Cheney in 2008, who wasn't in the best of health and was polling as poorly as George W. Bush was at the time. Nevertheless, Al Gore ran in 2000 after eight years with Bill Clinton. George H. W. Bush ran after two terms with Ronald Reagan. Richard Nixon ran in 1960 after serving eight years under Dwight Eisenhower.

But book after book and report after report said that behind the scenes, Obama *always* saw Hillary as the one to pass the presidential mantle to, as the right person to carry on his legacy:

First Black president.
First female president.
Ivy League law school guy.
Ivy League law school gal.
Perfect.

Biden, now seventy-three and losing a step, would be persuaded not to run in 2016, which (again) should have eliminated any thought of a third run for the presidency.

Somehow it didn't. Hillary decided not to make another go at it in 2020. Moreover, the other Democratic candidates were profoundly flawed, inexperienced, or simply unhinged. This would be Joe Biden's final run for a job he had believed should be his more than three decades ago.

Thanks to Donald Trump, COVID-19, and a media finally ready to embrace him, Biden would at last get his first real shot.

11

CIVIL RIGHTS CHAMPION JOE

The supposed secret to Joe Biden's political skill is his talent as a folksy storyteller. Sleepy Joe can persuade people that he's just like them. So the legend goes. Well, Joe's certainly a skilled storyteller, but it's tall tales he's spinning, not profound truths. One of his favorite tactics is to inflate his civil rights credentials, an exaggeration probably meant to cover for his dozens of Old White Guy gaffes.

For instance, remember the time he explained to a predominantly Black audience in southern Virginia that then–Republican presidential nominee Mitt Romney "said in the first hundred days he's going to let the big banks once again write their own rules. Unchain Wall Street! They're going to put y'all back in chains."

That was then–vice president Biden calling Romney and the latter's vice presidential candidate, Paul Ryan, the next coming of slave owners in 2012. Right on cue, instead of universal condemnation from the press, Republicans who were rightly outraged that a sitting vice president would say such a thing about two men with no racial animus in their respective histories got the "REPUBLICANS SEIZE!" treatment instead. That treatment, which can also include "REPUBLICANS POUNCE!" is defined by the Urban Dictionary thus:

> A headline in a newspaper or other article that describes Republicans (or other right-leaning individuals) attacking a Democrat (or other left-leaning individual) when that

> Democrat commits a misdeed. Always written by a reporter with left-wing political views, it will attempt to frame the Republicans as overzealous, and will either downplay, ignore, or excuse the Democrat's misdeed. Commonly done by the New York Times or Washington Post, it is often viewed as a sign of the bias within the media.

Anyway, "seize" was applied here. The *Los Angeles Times* announced: "Republicans seize on Biden's 'put you back in chains' comment."

"The Romney campaign seized upon the remark to argue that President Obama's campaign was trying to divide Americans and demonize his opponent," the report reads before adding this BS: "Though they did not touch upon the implication of slavery in making such a remark in front of an audience that included many African Americans."

For the love of God. Other headlines were similar.

> *Politico*: "Biden Draws Romney's Ire with 'Chains' Comment"
> *Washington Post*: "Biden: Romney's Approach to Financial Regulation Will 'Put Y'All Back in Chains' "

Here's more from the *Post*, which wrote that Biden "spoke before what appeared to be a racially varied audience of 900 people, and one prominent Republican suggested that his language could be interpreted as racially divisive."

Interpreted as racially divisive? Okay, sure. So only *Republicans* found this language offensive?

Joe Biden truly is the Joy Reid of Democratic politics. You've heard of Reid, an MSNBC prime-time host who is the very worst that cable news or television in general has to offer. This is the same person who referred to Supreme Court justice Clarence Thomas as "Uncle Clarence," a clear reference to Uncle Tom. That should have been a fireable offense. This is the same person who said Florida governor Ron DeSantis is rooting for COVID deaths of children in his state. This is the same person who said white supremacists elected Glenn Youngkin governor and Winsome Sears (who at last check is Black) lieutenant governor. That election took place in Virginia, a state Joe Biden won by double digits just one year before apparently it was taken over by the Klan.

It was Reid who slammed the media for covering the Russian invasion of Ukraine because (checks notes) only white people were involved. NBC News covered the war extensively, which makes Reid's coworkers at the mothership also secretly racist, perhaps without even knowing it. Or something.

The list goes on. Oh, by the way, Reid is also the same host who once wrote homophobic and anti-Semitic comments in her blog more than a decade ago. Instead of owning these hateful statements and perspectives, she blamed hackers for placing the content in question on said blog, and even went so far to notify the hack to the FBI.

"We have received confirmation the FBI has opened an investigation into potential criminal activities surrounding several online accounts, including personal email and blog accounts, belonging to Joy-Ann Reid," her attorney, John Reichman, said in April 2018.

Turns out no investigation was needed. A few days after claiming she was hacked, Reid went on the weekend MSNBC program she was hosting at the time to confess. Sort of.

"Many of you have seen these blog posts circulating online and on social media. Many of them are homophobic, discriminatory and outright weird and hateful. I spent a lot of time trying to make sense of these posts. I hired cybersecurity experts to see if somebody had manipulated my words or my former blog, and the reality is they have not been able to prove it," she explained.

"I have not been exempt from being dumb or cruel or hurtful to the very people I want to advocate for. I own that. I get it. And for that I am truly, truly sorry," she later added.

Several independent investigations into Reid's claim found that her blog wasn't hacked and the words were hers. Any halfway credible news organization would have fired her on the spot for doing so. But somehow, nothing came of this except Reid being *promoted* by MSNBC and awarded a prime-time show in 2020.

For his part, just like Joy, Joe Biden is also a classic example of someone saying racist, divisive, unhinged things and still getting rewarded by failing upward.

Exhibit A for this chapter: On a virtual appearance with the popular Black radio host Charlamagne tha God in 2020, Biden tried to end an interview before the host was ready, citing his wife's need to use their home television studio in Wilmington, Delaware.

"We've got more questions," Charlamagne informed Biden as he tried to exit.

"You've got more questions?" Biden responded. "Well I tell you what, if you have a problem figuring out whether you're for me or Trump, then you ain't Black!"

Think about what Biden was saying here to the Black community: If you don't toe the line, if you don't conform, if you don't vote for me, then you're like that Clarence Thomas guy. An Uncle Tom! Find a way any politician could sound more condescending and entitled to a voting bloc of millions, win valuable prizes.

Charlamagne, who is an excellent interviewer and host because he's both fearless and authentic, didn't accept Biden's answer.

"It don't have nothing to do with Trump. It has to do with the fact that I want something for my community," he explained.

"Take a look at my record, man," Biden said in response. "I extended the Voting Rights Act twenty-five years. I have a record that is second to none. The NAACP has endorsed me every time I've run. Come on, take a look at my record."

The NAACP doesn't endorse candidates.

"We want to clarify that the NAACP is a nonpartisan organization and does not endorse candidates for political office at any level," the NAACP said in a statement shortly after the interview.

It's also worth noting that Donald Trump actually performed better among Black voters than any recent Republican candidate: John McCain received just 4 percent of the Black vote in 2008. Mitt Romney captured just 6 percent in 2012. Trump got 8 percent in 2016 and 12 percent in 2020, or triple that of McCain and double that of Romney despite being called a racist on CNN and MSNBC basically on an hourly basis for four years.

During another presidential campaign, this time in 2020, CBS correspondent Errol Barnett (who is Black) asked Biden if he would take a cognitive test, which was (and is) a perfectly legitimate question. A CNBC/Change Research poll at the time found that 52 percent of likely voters believed that Biden was unfit to be president. Those numbers have only risen since he took charge and the spotlight has become brighter.

"No, I haven't taken a test. Why the hell would I take a test? Come on, man," Biden shot back before oddly adding, "That's like saying to you, before you got on this program, if you had taken a test were you taking cocaine or not. What do you think, huh? Are you a junkie?"

WTF? By all means, try the fun exercise at home in imagining what the reaction would be if *Trump* ever made a statement like that to a Black journalist. But perhaps the best part about Biden and the coverage of him is this notion that he's like some kind of white combination of Martin Luther King Jr. and Muhammad Ali. This was most evident when he attempted to claim that he was arrested in South Africa as he fought to see Nelson Mandela.

> **February 11, 2020:** "This day, 30 years ago, Nelson Mandela walked out of prison and entered into discussions about apartheid. I had the great honor of meeting him. I had the great honor of being arrested with our U.N. ambassador on the streets of Soweto trying to get to see him on Robbens [*sic*] Island."

> **February 16, 2020:** "After he [Mandela] got free and became president, he came to Washington and came to my office. He threw his arms around me and said, 'I want to say thank you.' I said, 'What are you thanking me for, Mr. President?' He said: 'You tried to see me. You got arrested trying to see me.'"

All. Of. This. Never. Happened. For if it did, one would think Biden's memoir would have mentioned something as monumental as a U.S. senator attempting to see the leader of the movement to end South African apartheid being arrested in a foreign country. The memoir never does. As for that U.S. ambassador who Biden says was with him, he told the *Washington Post* in 2020 there was "no chance" he "ever was arrested in South Africa."

"I don't think Joe was, either," he added.

Biden would eventually acknowledge he wasn't arrested when asked directly about it.

"I wasn't arrested, I was stopped. I was not able to move where I wanted to go," he explained on February 28, 2020.

Yup. Totally the same thing.

So why would Biden tell such a tale during a presidential campaign? All one has to do for an answer is look at how he would benefit from doing so, because timing is everything. The Biden-Mandela arrest story occurred heading into the 2020 South Carolina primary. Biden had already taken a beating in Iowa (fourth-place finish) and New Hampshire (fifth place). A loss in South Carolina would finish Biden off for good, marking yet another presidential campaign ending in humiliating fashion.

So Biden did the two things he does best:

a) Lie
b) Pander

The Mandela story kicked things off in painting himself as the best choice for Black voters in South Carolina. Senator Kamala Harris (D-CA) and Senator Cory Booker (D-NJ), the two most visible Black candidates, had already dropped out, leaving the field with basically the racial version of the cast of *Friends*. Biden's key to winning the Palmetto State rested solely on the endorsement of Representative Jim Clyburn (D-SC), which he secured on February 26 after promising to publicly announce he would choose a female Black Supreme Court justice if given the opportunity.

Two days later, after securing what he needed, Biden admitted the Mandela story was bullshit. One day later, he won the South Carolina primary, which turned around his entire campaign ahead of Super Tuesday. Exploiting race, even Nelson Mandela, a true leader, as a means to an end? For Joe Biden, this was his bread and butter.

As president, Biden continued the exploitation.

On January 11, 2022, during a speech on his "voting rights" package, which had been stalled in the Senate by two members of his own party (Joe Manchin and Kyrsten Sinema), Biden said: "I did not walk in the shoes of generations of students who walked these grounds. But I walked other grounds. Because I'm so damn old, I was there as well. You think I'm kidding, man. It seems like yesterday the first time I got arrested."

This. Also. Never. Happened. The *Washington Post*, to its credit, fact-checked this statement and gave it four Pinocchios, the worst rating it can give on the truth-telling scale. There is no fifth Pinocchio.

This kind of lying got especially sick—there's no other word for it—when Biden evoked his deceased mother to tell another story about being arrested for standing up for civil rights by . . . standing on someone's porch?

November 19, 2017, Nashville: "My mom, God love her, she sat there the whole time and my mom, after my dad passed away, we convinced her to move in with us and she was reluctant to do and anyway she was sitting there and I turned to her and said, 'Honey you haven't said anything' and she said, 'Joey'—it's a true story—she said, 'Let me get this straight, remember when you were fourteen years old and the real estate guys sold the house to a Black couple in an area—in a neighboring development called Graylyn Crest, I mean Carrcroft' and I said, 'Yeah, Mom.' 'Remember when I told you not to go down there, honey, because everybody is protesting and you got arrested standing with the family on the porch,' and I said, 'Yeah, Mom.' True story."

According to the *Post*, Biden told this story publicly on at least five occasions, with his age at the time changing in three different versions. But the thing that stands out about Biden's fibs is that they *aren't even that good*. So we're supposed to believe that a teenage white kid *got arrested* for standing on the porch of a Black couple? On what charge? *And* during a protest when police resources were already stretched thin?

So again, let's go to motive: Why would Biden choose to tell this story in 2022? Because he needed to sway public opinion on his voting rights bill. The bill was floundering after he had lied that Georgia limited voting hours in such a way that they didn't allow people enough time after work to vote. Again, a total lie.

"You're going to close a polling place at five o'clock when working people just get off? This is all about keeping workin' folks and ordinary folks that I grew up with from being able to vote," Biden said in an interview with (checks notes) ESPN, of all places, in 2021.

No. The new Georgia law still has its polls open until 7 P.M., which is the standard across most states in the country.

But this claim around "voter suppression" really gets me. I'm not a terribly emotional person, but this got me profoundly angry.

"When the Bible teaches us to feed the hungry and give water to the thirsty, the new Georgia law actually makes it illegal—think of this—I mean, it's 2020, and now '22, going into that election—it makes it illegal to bring your neighbors, your fellow voters food or water while they wait in line to vote," Biden said in January 2022 to students in Atlanta.

He was spinning a tale of an America where disenfranchised voters were fainting from hunger and thirst. Voting, in this America, is taking your *very life* into your hands. What the hell was he talking about? Could he be mixing American voting up with communist breadlines? It's incredible that anyone believed this tripe, but Biden's tall tales went over beautifully with a suggestive crowd. Having established his baseline lie, he riffed on it somberly, to great effect.

"What in the hell—heck are we talking about? [Applause] I mean, think about it! [More applause] That's not America! That's what it looks like when they suppress the right to vote!"

So let's get this straight, using this thing called logic. Nearly *every* U.S. state does not allow what's called "electioneering" near polling places, which includes political organizations and activists attempting to influence votes by bringing out food and water to voters otherwise helpless to stop off at a Kwik-E-Mart to pick up a bottle or snack along the way.

In Biden's home state of Delaware, for example, electioneering is banned within 50 feet. Remember, he served as a senator in Delaware for decades and therefore could have done something about this Jim Crow 2.0 thing rearing its ugly head in *his* state, but for some reason it wasn't a priority then. In Chuck Schumer's New York, electioneering is banned within 100 feet. In Georgia, the ban is placed at 150 feet. So we're talking *fifty feet*, which isn't even the distance from a pitcher's mound to home plate, in terms of the difference to the voting utopia states that are New York and California. Yet Georgia is the new Jim Crow 2.0.

Overall, it's actually easier to vote in Georgia than in New York, according to an analysis by the nonprofit, nonpartisan Center for Election Innovation & Research. As for Black turnout in Georgia, 64 percent of the state's eligible Black voters voted in the 2020 presidential election. In deep blue Massachusetts, for comparison, that number was just 36 percent, according to the U.S. Census Bureau.

It's also noteworthy that Biden made this pitch in January 2022 while in Georgia along with Vice President Harris. This so-called issue is a big deal to those like Stacey Abrams, the queen of "voting rights" since, she says, her election was stolen from her in 2018.

But just to underscore how toxic the president and vice president were entering their second year of office, when the president and vice president arrived in Georgia, Abrams, who is running again for governor, weirdly cited a scheduling conflict as the reason she couldn't appear with the two at a campaign event. We haven't seen a snub like this since *The Shawshank Redemption* didn't win Best Picture in 1994.

All this pandering to the Black community hasn't been working, either. Fewer than 7 in 10 Black voters (69 percent) said they supported Biden just fourteen months after taking office. Context: *More than 9 in 10* Black voters (92 percent) voted for him in 2020. This is a *23-point drop* in a relatively short period of time.

Corbin Trent, cofounder of the progressive No Excuses PAC, was blunt in an interview with *Politico* in January 2022 regarding Abrams finding something better to do during the president's and vice president's big visit. He said, "[Biden's] deeply unpopular. He's old as shit. He's largely been ineffective, unless we're counting judges or whatever the hell inside-baseball scorecard we're using. And I think he'll probably get demolished in the midterms. People will smell opportunity, and DC is filled with people who want to be president."

Trent is a former communications director for Alexandria Ocasio-Cortez.

Overall, the hill that Biden had decided to die on to begin his second year in office was "voting rights," with the president's speechwriters portraying any citizen who supported voter identification as nearly indistinguishable from Birmingham, Alabama's longtime police chief Bull Connor and Confederate president Jefferson Davis.

"Do you want to be on the side of Dr. [Martin Luther] King or George Wallace? Do you want to be on the side of John Lewis or Bull Connor? Do you want to be on the side of Abraham Lincoln or Jefferson Davis?!" asked Biden during that visit to Georgia.

That must make Elizabeth Warren on the side of Bull and Jefferson. Because early voting in Massachusetts, for example, is just eleven days. In Georgia under its new law, it's expanded to *seventeen days*. No matter. Warren called the changes in the Georgia law "a despicable voter suppression bill" as Democrats desperately attempted (and thankfully failed) to federalize voting laws in Biden's first year. By the way, Delaware, up until this year, had *no* in-person early voting. Period.

It's amazing that the Unifier in Chief decided to make this his first big issue in 2022 while his approval numbers were nosediving. Because if looking for one of the few issues that Americans are united on, requiring an ID to vote is one of them. Poll after poll indicates that an overwhelming share of the country supports having to show an ID in order to vote, including a majority of Democratic voters. A Monmouth 2021 poll, for example, finds more than 8 in 10 Americans support voters being required to show ID, including nearly two-thirds of Democrats. Other polls put support for requiring voter ID in the 70–80 percent range.

No matter. Enter the vice president, who actually made this hilarious out-of-touch argument against voter ID during an interview with Black Entertainment Television in July 2021.

"[To show proof of ID] you're going to have to Xerox or photocopy your ID to send it in to prove you are who you are," Harris told Soledad O'Brien when asked if there was room for compromise on voter ID laws. "Well, there are a whole lot of people, especially people who live in rural communities, who don't . . . there's no Kinkos, there's no OfficeMax near them."

Hold the phone. We live in the most advanced country in the world, and the argument is that some people don't have access to copy machines, which have been around since the 1950s? "People have to understand when we're talking about voter ID laws, be clear about who you have in mind and what would be required of them to prove who they are," she continued. "Of course people have to prove who they are, but not in a way that makes it almost impossible for them to prove who they are."

Almost impossible? What year—no, what *century*—does Harris think the rest of the country lives in? By the way, if voter suppression was truly real in any tangible way, doesn't anyone think by now that CNN or MSNBC would be doing town hall after town hall packed with those who are disenfranchised telling their harrowing tales of dehydration or hunger pangs while waiting in line to vote? Or how they decided they *really* needed to cast a drive-through vote at 3 A.M. and it was closed? Know this: If any remotely responsible and coherent person wants to vote in any election in this country, there are few things in life that are easier to do. If that wasn't the case, rest assured our activist media would be all over it with actual examples.

Donald Trump was elected in 2016 and earned 74 million votes in 2020 because his supporters knew they were electing a president, not a priest. It's why Trump connects with working-class voters so well despite being a New York billionaire. He doesn't speak down to those of color and doesn't make up shit about getting arrested in the name of civil rights, unlike Biden.

The Democrats' version of "voting rights" would ultimately fail in the Senate, 48–52. The president did everything in his power to get it passed. He accused fellow Democrats like Manchin and Sinema, and his Republican friends like Mitch McConnell, of siding with Bull Connor and Jefferson Davis. But when asked about his comments by RealClearPolitics' White House correspondent Phil Wegmann, Biden denied it happened, which is par for the course.

"Look what I said! Go back and read what I said and tell me if you think I called anyone who voted on the side of position taken by Bull Connor that they were Bull Connor!" Biden angrily responded. "That is an interesting reading of English. I assume you got into journalism because you like to write," he added.

Playing the tough guy. Joe Biden has been doing that going back to his days as a lifeguard when he allegedly stared down a gang leader named Corn Pop. And *that's* another story . . .

12

SNAP. CRACKLE. CORN POP.

The Biden administration has been a cockamamie catastrophe, but what's to blame? Is it just the same effects of liberal policy that we've seen before? Or does it lie in the answer to another question, one that many Republicans and Democrats and everyone in between are increasingly asking: Is the president of the United States mentally sharp?

He's the oldest president in U.S. history. He will be eighty in November 2022. A majority of the American people don't believe he has the mental chops to do the job. It's obvious: Joe Biden isn't the same person he was when he served as vice president just a few years ago. He certainly isn't the guy we saw at the vice presidential debate in 2012, when, while contemptuous and disrespectful to Paul Ryan, he showed a stiff rhetorical jab and energy.

People age differently in this world. My dad is seventeen days younger than Biden is and yet he enjoys a more active social life than I do. Alan Concha Sr., who worked six days a week as an accountant-turned-CFO at a printing company located in the shadow of Giants Stadium, earned every penny he made over a long career, and is now seemingly always out to dinner, seemingly always on vacation (he drove to Naples, Florida, from New Jersey in March 2022 just because he felt like it). He takes a daily jog around Packanack Lake in my hometown of Wayne, New Jersey. At seventy-nine, he still works as a part-time consultant to fill some hours, which allows him to dictate a routine at a pace he's comfortable with.

If private citizen Biden was losing his edge today, you wouldn't hear a *peep* from anyone about it. But we're talking about Commander in Chief Biden. Leader of the Free World Biden. And yet this Biden's schedule is lighter than my dad's on most days.

As an example, here's a story from December 2021. Omicron was raging. Testing was almost nonexistent. An outbreak of COVID had hit the White House.

"Has there been an outbreak of COVID at the White House, the NSC, the State Department, and at the Treasury? And has the president been in close contact with a COVID-positive person and thus in need of quarantine?" *New York Times* reporter Michael Shear asked Jen Psaki during her daily press briefing.

"The president has a full schedule today and is not in need of quarantine."

Psaki's vague answer caught my attention. On a hunch, I took a look at the president's schedule for that day. Here's what it looked like:

9:15 AM—The President arrives at the White House
10:10 AM—The President receives the President's Daily Brief
12:45 PM—The President has lunch with the Vice President
1:00 PM—Press Briefing by Press Secretary Jen Psaki
1:30 PM–The President meets with members of the COVID-19 Response Team on the latest developments related to the Omicron variant

That was it. Note: This isn't a cherry-picking exercise to push a narrative. Far too often, this is what Biden's schedule looks like. Late starts, early finishes. Very little travel.

But don't mistake Biden's light schedule for the first few innings of being senile. By and large, he's still running the ship, even if all cognitive pistons don't appear to be firing. The difference now is that he's making bad decisions that are reflexively liberal to satisfy his supporters both in terms of far-left voters and the media.

We saw this work ethic coming during the campaign, when, in the homestretch, Biden and his team called a lid on the day often before lunch and sometimes even before brunch time.

Here's what I wrote on September 26, 2020, on Twitter. Now, remember, this is less than six weeks before Election Day.

> Joe Biden called a lid on the day earlier this morning. No events, no questions. Less than 6 weeks before the election, the Democratic nominee is working at a 50% clip, calling a lid 7 of the past 14 days and 11 of the past 26 this month. Is this a preview of a Biden presidency?

Politico observed the same thing on September 26, 2020, but pooh-poohed it as only concerning to unserious people.

> Joe Biden called a lid—the arcane term reporters use when a politician is done traveling for the day—at 1:02 P.M. Friday; 9:20 A.M. Thursday; 9:22 A.M. Tuesday and 8:34 A.M. Saturday. Biden's early turn-ins provoked the Twitter set, professionally nervous Democrats (OK, bedwetters) and, notably, Donald Trump to scratch their heads or rain down ridicule.

But instead of criticism or concern about someone seeking the presidency calling it a day several hours before Fred Flintstone would ever dream of, *Politico* applauded not working as a genius strategy: "The Democratic nominee is sticking to his strategy: Keep a low profile, and let Trump light himself on fire."

Funny how that "strategy" has extended well into his presidency, when he's supposed to be working for the American people 24/7. In Biden's world, that perhaps means twenty-four hours a week, seven months per year. So how is this strategy being received by voters? Well . . . a March 2022 poll by ABC News and the *Washington Post* found that 54 percent of Americans didn't believe that Biden had "the mental sharpness it takes to serve effectively as president." Imagine that: A majority of the greatest country in the world doesn't believe its leader is mentally sharp.

A February 2022 ABC News/*Washington Post* poll also found that 59 percent said they do not think Biden is a strong leader.

This sentiment matches what many Trump supporters and those actually objective in the media were warning about during the campaign. There was a reason (and no, it wasn't because of COVID fears) that Biden's handlers largely kept him in the basement while running for president. It was because they could see privately what they didn't want voters to see publicly.

Here were some of candidate Biden's not-so-greatest hits during the campaign.

In August 2019:

> "One of his buddies got shot, fell down a ravine about 60 feet. A four-star general asked me whether I'd go up in the [Forward Operating Base]," Biden told a modest crowd at Dartmouth College about a visit he made as vice president to Afghanistan's Kunar Province in 2011. "And everybody got concerned [that] a vice president [was] going up in the middle of this, but we can lose a vice president, we can't lose many more of these kids, not a joke. This guy climbed down a ravine. Carried his guy up on his back under fire, and the general wanted me to pin the Silver Star on him."

"God's truth, my word as a Biden," the eventual Democratic nominee continued. "He stood at attention, I went to pin him, he said: 'Sir, I don't want the damn thing. Do not pin it on me, sir, please. Do not do that. He died. He died.'"

Moving story. And it must be true because Joe said it was God's truth and his word as a Biden.

Turns out much of it wasn't true. From ABC News: "Biden visited Kunar province as a U.S. senator in 2008, not as vice president, and the soldier involved was not an older Navy captain, but a 20-year-old Army specialist named Kyle White. White was awarded the Medal of Honor in 2014, and Biden was present at a White House ceremony when then-President Obama pinned it on his chest."

So did Biden outright lie or was he confused? Could it be both? While there was some reporting on this, there were no angry op-eds in the *New York Times* or *Washington Post* roasting Biden for pushing a story to make him look brave and magnanimous that involved a dead U.S. military member. Is it the result of cognitive decline? According to Rasmussen Reports, two-thirds of voters want Biden to take a cognitive test. That perception speaks volumes and indicates that more and more Americans don't believe he's all there now.

But the most fascinating tale Biden told on the campaign trail (to end all tales) can be summed up in two words:

Corn.

Pop.

It was the summer of 1962. Joe Biden was twenty. Life was especially good after the future president took a job as a lifeguard in Wilmington, Delaware, while home from school. The swimming facility was located in a predominantly Black area of town. Biden, in his memoir, said he took the job "in hopes of learning more about the Black community."

Uh-huh.

Biden also says he was popular at the pool because many of the Black people there (checks notes again) "had literally never talked to a white person." In a related story, according to 1960 census figures, Wilmington was still *73 percent white.*

The pool where Biden worked was, according to his memoir, home to a gang called "The Romans." Who was the gang leader? A guy named Corn Pop (this is Wilmington and not Watts, after all).

Well, anyway, one day Corn Pop did what all badass gang leaders do: He and other members of the Romans decided they wanted to play Sharks and Minnows and perhaps a few games of Marco Polo. But Corn Pop violated the rules regarding (checks notes) the diving board, which is the equivalent of committing a drive-by shooting in Delaware.

Lifeguard Biden, the David Hasselhoff of in-ground pools, could not and *would not* let this stand, so he expelled Corn Pop from the pool and was a profound dick in the process of doing so.

"You! Off the board! Or I'll come up and drag you off!" Biden said he shouted before also referring to the gang leader as "Esther," which was apparently a reference to swimmer Esther Williams, who was a thing at the time. Mr. Pop also apparently refused to wear a bathing cap, drawing Biden's ire.

But think about this for a moment: Why would Biden go from Hasselhoff to full Clint Eastwood in instigating a confrontation with Corn Pop (again, over a bathing cap and diving board etiquette) when he could have just respectfully asked him to put one on and made him promise not to do any cannonballs?

But the Pop wasn't no pussycat. He proceeded to confront young Joe Biden in the parking lot.

"He was waiting there with three guys [with] straight razors," according to Biden, who armed himself with a chain that was used to separate the low end from the deep end in the pool.

Wow. What kind of lifeguard training could prepare him for this? The future chairman and ranking member of the Senate Armed Services Committee was about to get shivved over swimwear!

Here's what Biden claims happened next:

"I said: 'First of all . . . when I tell you to get off the board, you get off the board, and I'll kick you out again. But I shouldn't have called you Esther Williams, I apologize. But I don't know if that apology is going to work."

"He said, 'you apologize to me?'" Biden continued. "I said, 'I apologize but not for throwing you out, but I apologize for what I said.'"

Corn Pop proceeded to back down, which is how things totally happen in Chicago or Los Angeles these days.

"He said OK, closed the straight razor and my heart began to beat again," Biden recalled.

To unpack this:

a) A gang leader is openly humiliated in front of other gang members by a lifeguard.
b) Said gang leader confronts the lifeguard.
c) Lifeguard reiterates that he would kick him off the diving board again if in the same situation.
d) Lifeguard apologizes for referring to the gang leader as a female swimmer and actress.

And then . . . the gang leader folds like a cheap bathing suit?

This is an impossible story to truly verify. Corn Pop, whose real name was William Morris, died in 2016. So, again, is this a story that Biden largely made up? Or is his mind not exactly all there? Or a combination of the two?

There are many odd stories and claims to choose from with Joe

Biden. Given that there's fifty years of material to work with, we could fill this whole book with them. The bizarre blather has obviously continued into the presidency, but Biden rarely gets called out on this stuff on any major level. Again, if Trump had said this stuff, we'd be hearing about the "Liar in Chief" and Trump's "alternative facts" in the most pious way possible, perpetually.

"I have known every [Israeli] prime minister well since Golda Meir, including Golda Meir," Biden claimed in December 2021. "And during the Six-Day War, I had an opportunity to—she invited me to come over because I was going to be the liaison between she and the Egyptians about the Suez. And I sat in front of her desk."

Wow. Impressive stuff. The Six-Day War between Israel and an Arab coalition of Jordan, Syria, and Egypt occurred just five years after Lifeguard Biden stared down an armed Corn Pop. Now here he was, sitting across from an Israeli prime minister, the iconic Golda Meir, the first female leader of Israel, advising her on the war!

The problem for Biden is that he was still a student at Syracuse Law School in 1967 when the war broke out and when this meeting supposedly had taken place. Oh, one more fun fact: Golda Meir didn't assume office until *1969.*

Biden ran for the presidency as a guy who would be George Washington 2.0 compared to Donald Trump. He promised to restore trust in the government and our leaders, and to unify the country. In his inaugural speech, he declared: "Recent weeks and months have taught us a painful lesson. There is truth and there are lies. Lies told for power and for profit. And each of us has a duty and a responsibility as citizens, as Americans, and especially as leaders. Leaders who are pledged to honor our Constitution, to protect our nation. To defend the truth and defeat the lies."

The media lapped it up. Because of course they did.

"Joe Biden's Love Letter to the Truth; The new President tried something different: levelling with the American people," swooned the *New Yorker* after the address.

"'Defend the Truth and Defeat the Lies': Biden Moves Past Trump's War on Media," bellowed taxpayer-funded NPR.

Rolling Stone lovingly intoned, "The Moment Met Joe Biden; Biden saw a country in peril but never lost faith in it, and that's why he's president today."

"It was the best inaugural address I have ever heard," then–Fox News anchor Chris Wallace concluded. Wallace would go on to CNN+, the network's streaming service, which folded in the span of thirty days.

One year later, polls show that Biden is less George Washington and more George Costanza when it comes to telling the truth. According to a CNN poll conducted by SSRS in December 2021, two-thirds of Americans say they doubted if Biden was "a leader you can trust." Before you say this is a number skewed by Republicans who will knock Biden on everything and anything, the number includes *75 percent of independents* and more than a third of the president's own party.

Look, mocking Biden on his unsteady grasp of the truth or reality is one thing, but there is a serious aspect to this regarding his "gaffes," because they can present serious national security issues.

Here's an example from a CNN town hall in October 2021. College student Glenn Niblo said to Biden, "China just tested a hypersonic missile. What will you do to keep up with them militarily and can you vow to protect Taiwan?"

Biden proceeded to insert his foot into his mouth. "I have had—I have spoken and spent more time with Xi Jinping than any other world leader has. That's why you have—you know, you hear people saying, Biden wants to start a new Cold War with China. I don't want a Cold War with China, I just want to make China understand that we are not going to step back, we are not going to change any of our views."

CNN's Anderson Cooper tried to draw a point out of the Biden rambling. "Are you saying that the United States would come to Taiwan's defense?"

Biden, brief for once, said plainly, "If China attacked? Yes, we have a commitment to do that."

So if China attacks Taiwan, the U.S. commander in chief is saying, we'll attack China militarily? Because that hasn't been our policy on this, like . . . ever.

Jen Psaki cleaned up on Aisle 5 the following day. By lying, of course.

"The president was not announcing any change in our policy nor has he made a decision to change our policy," Psaki said despite her boss saying quite the opposite on national television the night before.

In other words, ignore what the old man said . . . we ain't sending in our military if China decides to pull a Russia and invade Taiwan.

These mental lapses also extend to other matters, such as COVID-19. During another CNN town hall, this time with the Cronkite of our time, Don Lemon, Biden was doing his usual attack on the unvaccinated when he made this claim.

"You're not going to get COVID if you have these vaccinations," he said.

Ah, okay. That's fascinating because my wife, an ER doctor who has been seeing COVID patients for more than two years, got vaccinated. Three times. She still got the Delta variant in November 2021 after my kids brought it home from a neighbor's house.

She wasn't alone, of course. Breakthrough cases were all the rage around Christmastime in 2021, even among those vaccinated and boosted.

But when faced with yet another ridiculous Biden gaffe, the pundits shrugged in the exact same way that John Madden used to talk about quarterback Brett Favre with Pat Summerall after Favre would panic and throw across his body while under duress into double coverage, resulting in an interception.

"Well, Pat, that's just Brett Favre being Brett Favre! It's like a kid drawing up plays in the dirt on a playground out there. Brett Favre is gonna make mistakes, no doubt about it. But sometimes you have to gamble to win big in the game of football. And in the game of life. And Brett Favre is a gambler."

Note: I *loved* John Madden as a coach and in the booth. Greatest of all time in the latter category. But that excuse was BS and I suspect he knew it, too.

Pundits always go into Madden mode when Joe Biden makes another gaffe. You can just hear the affectionate shrug in the tone they use describing the most powerful man in the world. Not exactly "truth to power," is it?

> *Politico:* "Joe Being Joe—The best Biden gaffes, slip-ups, uncomfortable truths and plain old bloopers"
>
> CNN: "It's Biden being Biden. Again."

New York Times pundit Paul Krugman wrote a puff piece called "In Praise of Fallible Leaders," in which he presented Joe's mistakes not as dangerous miscalculations from an American president, but rather as charming faux pas that give our Beloved Leader the opportunity to demonstrate his humble nature.

> Biden . . . while he may not be the most impressive presidential candidate ever, is clearly a man comfortable in his own skin. He knows who he is, which is why he has been able to reconcile with former critics like Elizabeth Warren. And when he makes a mistake, he isn't afraid to admit it. Over the past few months we've seen just how much damage a president [Trump] who's never wrong can do. Wouldn't it be a relief to have the White House occupied by someone who isn't infallible?

So Biden will always admit mistakes when he's wrong, huh? In *Jeopardy!* parlance, I'll take "Afghanistan withdrawal" for $500.

Sharply contrasting with Krugman's idea of Biden, the real Biden gutlessly pointed the finger at everyone else. On August 31, Biden didn't admit mistakes around the fall of Kabul, which happened far earlier than predicted by U.S. intelligence and his military advisers. Instead, he blamed the Afghan army, the same one he bragged about being so fully trained and ready just one month prior. And, as this administration *loves* to do, Biden also blamed Trump for forcing him to pull out of Afghanistan.

"When I became president, I faced a choice—follow through on the [Trump] deal [with the Taliban], with a brief extension to get our forces and our allies' forces out safely, or ramp up our presence and send more American troops to fight once again in another country's civil conflict," Biden said in a thirty-minute address from the White House on August 31, 2021.

He just didn't have any choice, see? He just had to do what Trump planned to do. Right. Since when? Remember, Biden signed dozens of executive orders in his first two weeks reversing Trump-era policies. In fact, Biden signed more executive orders than Trump, Obama, or Bush in his first one hundred days. Yet we're supposed to believe that Biden *had* to abide by a deal his hated predecessor made? He could have just renegotiated another deal with the Taliban while calling the deal Trump made moot.

"The bottom line is, there is no evacuation from the end of a war that you can run without the kinds of complexities, challenges, threats we faced. None," Biden insisted. The president would go on to call the U.S. evacuation an "extraordinary success." There was no mention of Biden's vow during interviews provided prewithdrawal to extract every U.S. citizen who wanted to leave.

When his address was over, reporters began to shout questions. Biden, as he has done so often throughout his relatively young presidency, turned his back and walked away without answering a single one. No column came from Krugman admitting he was wrong in declaring that Biden would admit mistakes as they occurred.

But unlike Krugman, most Americans aren't wearing rose-colored spectacles. Two-thirds of American people want the president's brain tested, according to a March 2022 Rasmussen poll mentioned earlier, including 43 percent of *Democrats*. In the same poll, 63 percent say they have seen a mental decline from Biden.

We have a president who will turn eighty this year, and a vice president who may only be in her fifties but consistently dumps incoherent word salad on the world stage when without the aid of a teleprompter. Speaker of the House Nancy Pelosi, who is in her eighties, increasingly acts and says things that are profoundly odd.

So why is it that the media shrugs these stories off? Why are they so willfully blind to the ridiculous tall tales of the left? I think I know the answer.

13

GET THIS MAN ELECTED

Complaining about the liberal media is one of the mainstays of pragmatic, commonsense punditry. With such a varied selection of imbeciles, lickspittles, and brownnosers to choose from, who could blame us? But for all the misleading stories that have been written about Joe Biden in the last few years, it's worth asking a question. Are journalists trying to fool us? Or are they fooling themselves? One thing's for sure: You'd have to be a fool not to see how skewed the news is now.

The bias is clear even to the least informed voter who hasn't lost their common sense. Way back when, in 1992, I first noticed media bias. It was during a presidential debate, which I was surprisingly watching instead of a sporting event or being out somewhere on campus. To that point, in my college years, I was more of a *SportsCenter* than C-SPAN guy.

At twenty-one, I was like any other guy at school and wasn't paying much attention to the political scene. Quite frankly, it was a bit tedious at the time. There was no Donald Trump, unless reading the *New York Post*'s gossip page ("Page Six") was your thing. And there wasn't the unintentional comedy that is Joe Biden and Kamala Harris to make us laugh and cringe simultaneously. Most of all, there was no Fox News or MSNBC, just a responsible, sane version of CNN.

But H. Ross Perot and Bill Clinton had made things interesting that election year and I suddenly found myself immersed in what looked to be a close race after three straight blowouts by Republicans Ronald Reagan (twice) and George H. W. Bush (once). The latter was seeking a second term, which would make it sixteen straight years of Republican rule in the Oval Office.

But a recession had hit and the sky-high approval ratings for Bush from the Gulf War had quickly evaporated. Now the incumbent was in serious trouble as he entered the second general election debate, moderated by Carole Simpson of ABC News.

Being a political media novice, I had zero idea who Carole Simpson even was. But like most Americans, I assumed that any journalist chosen to moderate a debate watched by 70 million people would be on the same objective level as CNN's Bernard Shaw or ABC's Peter Jennings, two respected anchors at the time.

I was wrong. The bias was subtle, but I specifically remember throwing an empty Bud Dry can at my twenty-five-inch TV, which weighed more than I did (and therefore could take the punishment), after Simpson showed which team she supported. No, I wasn't dismayed because I was some kind of Bush superfan, by the way. It just suddenly felt like he had three opponents on that stage: Clinton, Perot, and the debate moderator herself.

The moment came about halfway through the debate, when Bush was asked a question from an audience member in a town hall setting, which is customary for the second of three presidential debates.

The question was *just a bit* hyperbolic.

The man in the audience said, "We've talked a lot tonight about creating jobs. But we have an awful lot of high school graduates who don't know how to read a ruler, who cannot fill out an application for a job. How can we create high-paying jobs with the education system we have and what would you do to change it?"

Hold the phone: An awful lot of high school graduates *can't read a ruler*? Who vetted this question and allowed it on the air? It's understood that this was a town hall setting, but the organizers can't just let anyone throw something like this out there without some kind of fact check around it, right?

But it happened. Then Simpson jumped in to decide which of the three candidates would get the question.

"Who would like to begin? *The education president?*" she said sweetly.

The dig was obvious. Bush had once said that he wanted to be known as the education president, and the implication from Simpson and the guy asking the question was that he was clearly failing America's children during his first term. An awful lot of those with high school diplomas couldn't even read a ruler, after all.

Perhaps someone could have brought up the fact that Arkansas, where Bill Clinton was governor for two terms, was ranked forty-sixth in the country on the education front. According to a story in the *New York Times* in August 1992, two months before this presidential debate:

> The Arkansas school system, ranked among the worst in the nation when Mr. Clinton first took office, is still near the bottom in most national ratings. And state officials acknowledge that real improvement is years away. And so Governor Clinton's aides acknowledge that there is no guarantee that his ideas will produce positive results nationally. Since Mr. Clinton took office in 1979, more Arkansans have gone to college. But at the same time, college-admittance test scores have fallen. The state still ranks near the bottom in teachers' pay, 46th among the 50 states and the District of Columbia. It ranks 48th in spending for each pupil.

Yet, given all of those data points to work with, Simpson's "question" for Clinton still only consisted of two words after Bush was done speaking.

"Governor Clinton?"

Yup, that was it. Nothing about his record as governor. No question *at all*, in fact. Just an open-ended invitation to the governor to take the ball and run untouched into the end zone.

Twenty years later, and much like CBS's Dan Rather did in exposing himself as a Democratic activist after he no longer occupied an anchor chair at a major broadcast network, Simpson exposed herself as the activist she was all along. Just without the subtlety.

If you recall, Rather, along with producer Mary Mapes, had attempted to make a serious impact on the 2004 presidential election (which was extremely tight) just days before Election Day. Rather, reporting for *60 Minutes*, had dropped a huge bombshell that seemed to show Bush lying about his military service in the Texas Air National Guard during the Vietnam War. But it was quickly revealed the whole foundation of the Rather and Mapes report was based on documents proven to be fabricated. Rather and Mapes were fired from CBS not long after. To this day, Rather, who gets invited to talk about media ethics on a CNN program called *Reliable Sources, insists* the report was accurate despite being based on fake documents. Mapes has defended the report as well.

The former *CBS Evening News* anchor would go on to lead the resistance against Trump largely through his social media screeds, which read like something out of a bad Aaron Sorkin movie.

"As I settle into a Saturday evening I breathe deeply, reflecting on how nice it feels to know that my sanity won't be tied to the destructive whims of Mr. Twitchy Twitter Thumb," the eighty-eight-year-old Rather tweeted after Joe Biden took office.

The tweet received nearly 78,000 likes.

For more examples, feel free to google "Dan Rather" and "Trump" if you have a few hours to kill in an airport sometime, because watching the recipient of the Edward R. Murrow Award for Lifetime Achievement morph into Keith Olbermann is one hell of a ride. (On the other hand, perhaps he was just living down to Murrow's example. It's not often mentioned now, but the legendary TV journalist got his first job at CBS by lying on his job application.)

For proof, here is a small collection of columns on CaroleSimpson.com. It might as well be DNC.com or MSNBC.com or AOC.com.

March 2012: "Fight Women, Reelect Obama"
Barefoot, pregnant and staying home. That's where Republican men apparently want to see women of the 21st Century.

November 2020: "2020 Election Dilemma"
Trump may be able to take some comfort in the fact that the race was close, very close. But that's what has me stumped. In this election, voter turnout was the highest it has ever been. 75 million votes for Biden, more than 70 million for Trump. How is that possible? Almost half the votes going to Trump? While everybody I know was celebrating the Biden win—there must have been nearly as many people angry and sad about Trump's loss.

January 2021: "Inauguration Jubilation"
I have something to say about the inauguration of President Joe Biden and Vice President Kamala Harris. Now I know once

again how it feels, to be happy. I haven't been happy for four years.

The blog reads no differently than one for your average Lincoln Project founder's diary. But Simpson didn't just suddenly decide to root for one side and vilify another. It's been baked into her DNA all along. She *wanted* Bill Clinton to win that debate in 1992 that night. She wouldn't have conducted herself the way she did if that wasn't the case.

There's also one very telling thought in the three blog excerpts above: "While everybody I know was celebrating Biden's win . . ."

Yep. Almost everyone in journalism *was* celebrating Biden's win. CNN literally had one of its prime-time anchors, Don Lemon, crying on the air.

"I almost can't talk right now, because of the emotion," Lemon told viewers. He wasn't alone. We hadn't seen this kind of blatant celebration since Dorothy's house fell on the Wicked Witch of the East.

Contrast the reactions to when the winner of the 2016 election was called as Donald Trump. The *New Yorker* called it "An American Tragedy." CNN's Van Jones, fighting back tears, declared Trump's win was a result of "a whitelash against a changing country. It was whitelash against a Black president in part. And that's the part where the pain comes." (Jones would cry tears of joy when Biden defeated Trump four years later.)

Or take NBC News' blaring headline: "12 DAYS THAT STUNNED A NATION. HOW HILLARY CLINTON LOST."

Let's return to our opening question. Who's getting fooled here? The reality is that a lot of journalists are living in a bubble of their own making. Our political media largely exists in either New York or Washington. Those cities are overwhelmingly blue. To survive in this business on any major national level (outside of Fox News, anyway), one must conform to a certain worldview; one must embrace certain policies and ideals. If that doesn't happen, as sure as you were born, you *will* be ostracized.

Bari Weiss, a left-of-center columnist formerly of the *New York Times*, said it best in her resignation letter from the paper two years ago.

"Twitter is not on the masthead of The New York Times. But Twitter has become its ultimate editor," she wrote in July 2020. She continued:

> As the ethics and mores of that platform have become those of the paper, the paper itself has increasingly become a kind of performance space. Stories are chosen and told in a way to satisfy the narrowest of audiences, rather than to allow a curious public to read about the world and then draw their own conclusions. I was always taught that journalists were charged with writing the first rough draft of history. Now, history itself is one more ephemeral thing molded to fit the needs of a predetermined narrative.

Amen.

Back to me. After bouncing around in sportswriting and working in IT sales to pay the bills (it ain't cheap living in or near New York City), I eventually entered the niche occupation of media reporter/columnist/analyst, beginning in 2012 courtesy of writing columns for the affable Dan Abrams and his unique offering, Mediaite, which is widely read within the cable news industry.

Four years and a ton of cable news appearances later on Fox News and CNN, I moved to the reporting side by joining *The*

Hill, one of the few down-the-middle political publications left in this business, on a full-time basis. Quality people. Professional. Friendly. Solid publication. A great experience. In 2020, upon signing with Fox News as a contributor, I jumped to the opinion side of *The Hill* and do two columns a week. In short, it's a labor of love doing both TV and writing and actually getting paid to do it.

Having covered the media industry for more than a decade, I can safely say that things have become much, *much worse* on the media bias front. The obvious turn from subtle to blatant bias occurred during the coronation of Barack Obama in 2008. The coverage was even more horrific during his reelection victory in 2012 over Mitt Romney, who somehow was portrayed not only as a racist, but also as a nonfictional version of Gordon Gekko (Michael Douglas) from the movie *Wall Street* and a guy who also didn't treat women or dogs very well. As we all witnessed, the 2016 election between Donald Trump and Hillary Clinton involved blatant bias in broad daylight that led "reporters" to conduct themselves in ways that, at one time, would have resulted in their firing.

Some in this media niche refer to me as a "right-wing" or "conservative" media guy. But it's not about ideology with me, or right versus left. It's about pragmatism. Common sense. Logic. It's also about journalists doing their jobs the way Tim Russert or Ted Koppel or Bernard Shaw or Mike Wallace did before them.

To that end, journalism, ladies and gentlemen, has now gone *way* past just plain old bias. We've passed into another *ism*, activism, and that jump was cemented the moment Donald Trump came down that escalator at Trump Tower in June 2015. This only accelerated when many in the media scolded themselves for not taking Trump seriously enough in 2016 while even propping him up for clicks and ratings leading up to Election Day. When

2020 came, they simply *were not* going to allow his reelection to happen.

His opponent could have been anybody. Didn't matter. Trump *had* to be stopped. The only problem was that not many bothered to explore exactly who would be replacing Trump. There was just no interest by most of the media in investigating Joe Biden's checkered past. We further explore the consequences of this media metamorphosis in this book. It truly is, for lack of a better word that doesn't exist, a *craptastrophe*.

The left-wing monoculture in the media is so strong that, in 2020, they just couldn't bring themselves to ask Joe Biden any tough questions. How could anyone attack our best chance at defeating Trump? It's dangerous—no, it's damn well *unpatriotic*.

Flash back to just after Labor Day 2020. Things were looking very good for the Democratic Party and its nominee to unseat Donald J. Trump. An ABC News/*Washington Post* poll had Joe Biden, who had never won a presidential primary before this year, up by 10 points nationally. In swing states from Florida to Pennsylvania to Nevada, he was also leading, albeit by smaller margins.

The amazing thing about this lead was that Biden was running for the highest office in the land by essentially pleading the Fifth. Press conferences were few and far between for Joe Biden. Just one in April 2020 via video, and one in person in Delaware in July. In that time, Trump was front and center on most days taking questions primarily on coronavirus for sixty, ninety, 120 minutes. Reporters would get multiple follow-ups. Very few questions were easy. Contrast that with Biden on the rare times when he did take questions, when most of the queries thrown his way were questions about Trump. There's no easier question for any candidate to field than those that ask to define one's opponent.

But one question from Fox News' Doug McKelway did prompt a

curious answer from Biden. McKelway said, "I'm sixty-five. I don't have the word recollection that I used to have. I forget my train of thought from time to time. You got twelve years on me, sir. Have you been tested for some degree of cognitive decline?"

"I've been tested, and I'm constantly tested," Biden replied. "I can hardly wait to compare my cognitive capability to the cognitive capability of the man I'm running against."

Joe Biden has been and is . . . *constantly tested*? By whom? A doctor? What's the doctor's name? What were the results of the tests?

In a normal year with a normal media, this would be the takeaway from the press conference. Cable news *Brady Bunch* panels would be ablaze with armchair psychiatrists (who aren't actually psychiatrists) attempting to analyze exactly what the guy seeking to be president meant. And (rightly so) the candidate would be met with questions during every interview about being constantly tested for cognitive decline.

Instead, we got headlines like this:

> *Politico*: "Scenes from Biden's First Encounter with the Media in Months; The Trump campaign took credit for smoking Biden out of his basement—only to see him give a gaffe-free performance."

Gaffe-free? Other headlines promoted the doddery Biden with the sort of action verbs more appropriate to Muhammad Ali.

> NBC News: "Joe Biden Rips Trump as Coronavirus Surges: 'The Wartime President Has Surrendered'"
>
> CNN: "Biden Slams Trump on Coronavirus: 'Our Wartime President Has Surrendered'"

Deadline: "Joe Biden Blasts Donald Trump's 'Cognitive Capability' & 'Dereliction of Duty'; Promises No Campaign Rallies in COVID-19 Pandemic"

Boy . . . If I didn't know better, I'd think that maybe these outlets were picking a side in their nonpartisan coverage of the race. But it was clear how this was lining up for Biden's handlers, and they couldn't have been more happy with the circumstances.

It was obvious Biden wasn't the man he was even a few years ago. The more voters saw of Biden, the less confident they would become in his ability to do the job. Months later when he was front and center on weekdays as the forty-sixth president, his poll numbers dropped, especially on leadership and competence and having the mental sharpness to do the job.

Meanwhile, Trump was facing a barrage of profoundly negative stories in addition to coronavirus, most notably this "bombshell" by the *New York Times*: "Russia Secretly Offered Afghan Militants Bounties to Kill U.S. Troops, Intelligence Says; The Trump administration has been deliberating for months about what to do about a stunning intelligence assessment."

The story, based on anonymous sources, as so many "bombshells" during the Trump era were, accused Trump of sitting on intelligence that the Russians had bounties out on U.S. troops. Here we go again: Trump is a Russian agent, and therefore allowed U.S. service members to be killed in order to protect his true boss, Vladimir Putin. The *Times* article went on:

> The intelligence finding was briefed to President Trump. . . . Officials developed a menu of potential options—starting with making a diplomatic complaint to Moscow and a demand that

> it stop, along with an escalating series of sanctions and other possible responses, but the White House has yet to authorize any step, the officials said.

Lo and behold, this *New York Times* bounties story was "confirmed" by the likes of the *Washington Post* and CNN. In other words, they depended on the same "sources" as the *Times* did. So was the story true? Because that's treason if Trump, the commander in chief, was told this and did nothing. Trump denied all of it.

"Intel just reported to me that they did not find this info credible, and therefore did not report it to me or @VP. Possibly another fabricated Russia Hoax, maybe by the Fake News @nytimesbooks, wanting to make Republicans look bad!!!" he tweeted at the time, when he still had a Twitter account.

No matter. CNN and MSNBC and the *Times* and the *Post* and the major broadcast networks all ran with the *Times* story as absolute fact. This gave Biden an easy salvo to fire at Trump repeatedly throughout the campaign, including in the debates in the fall.

"I don't understand why this president is unwilling to take on Putin when he's actually paying bounties to kill American soldiers in Afghanistan," Biden said of Trump in front of 80 million people on October 22.

Of course, the story wasn't true, but we didn't find that out until April 2021, when Biden was comfortably installed in the White House. On cue per the BBC, "The White House has acknowledged there was little evidence that Russia had offered Taliban militants bounties to kill US soldiers in Afghanistan. A spokeswoman for President Joe Biden said the claim had 'low to moderate confidence' from U.S. spy chiefs. Russia has denied paying the bounties. In last year's US election, Mr. Biden heavily criticised Donald Trump for not confronting Russia over the claim."

Oh.

It all harks back to the 2012 campaign, when then–Senate majority leader Harry Reid accused Republican nominee Mitt Romney of not paying his taxes for ten years. Some in the media questioned the claim but more than a few ran breathlessly with the nutty allegation, because Romney needed to be stopped and Obama-Biden *had* to be reelected.

"The word's out that he hasn't paid any taxes for ten years. Let him prove that he has paid taxes, because he hasn't," Reid said.

When asked about the dubious claim by CNN years later, after it was clear it was total BS, Reid patted himself on the back.

"Romney didn't win, did he?" he proudly responded.

Biden didn't directly do what Reid did, but one has to ask: How did so many outlets, including the *New York Times*, *Washington Post*, and CNN, among others, *all* get the story wrong? Why, almost like clockwork, do these "mistakes" go in one direction against conservatives or Republicans? When have we seen, during the entire Biden campaign and presidency, one example of a massive media "mistake" going against liberals or Democrats?

Because maybe, just *maybe*, these "reporters" see and hear what they want to see. If they're fed unverifiable information from a questionable source, their eyes get bigger than their stomach and they swallow the information whole. In an age when getting it first trumps getting it right, this is what happens.

So, was anyone fired or even disciplined over the Russian bounty story seeing the light of day? Of course not. Mission accomplished. Trump had this allegation wrapped around his neck for months leading up to Election Day. Only in April 2021 did U.S. intelligence magically conclude that there was nothing to the story.

The Swamp is very real. And very deep. And having a media play along so easily keeps the misinformation machine humming.

Many in media are like seagulls at the beach. Throw anything up into the air, and they will absorb it whole without much thought on what they are absorbing. Dubious sources exploited these reporters time and again. They knew verification on explosive information had largely gone out the window because those being fed really, really wanted to believe the Russians altered the 2016 election or the Russians were paying to have U.S. service members killed and Trump was being blackmailed by the Kremlin to stand down.

As for Reid and Romney, ask yourself what you would do if you were Mitt. I know what I would say, and it wouldn't be pretty. In a 2021 interview shortly before his death, Reid revealed that Romney had agreed to meet.

"We had a meeting at his home in Salt Lake where Mitt and I talked about probably how we had done things wrong about each other and our wives were there and we had a very nice meeting—kind of a make-up session," Reid shared. "So, I admire Mitt Romney. I think he's a very very fine human being."

Grow a pair, Mitt. Geez. How can anyone forgive someone for doing such a thing? Ah, right: Mitt Romney would.

Back to 2020: Biden's handlers understood that they had to put the candidate out there periodically in settings that played to his alleged strengths.

What better way to do that than with town halls arranged by friendly news organizations that would make *absolutely sure* Mr. Biden was never made to feel uncomfortable or challenged in any way?

14

MOSTLY PEACEFUL RIOTS

When you look back on the way violent crime and riots and looting and the chaos in general were covered in 2020, it's hard to overstate just how partisanly many in our media conducted themselves. It was (and still is) several flavors of stupid.

Stupid. There really isn't a better way to describe the summer of 2020 and the way it played out on our television screens. Here we had a presidential election coming up. We also had a pandemic shutting down the country. Then the police killing of George Floyd happened in Minneapolis, which resulted in outrage and mayhem in cities big and small across the country, like Kenosha, Wisconsin.

It was there that the police shooting of Jacob Blake occurred in August 2020. Blake, wanted for a felony warrant, was armed with a knife and had resisted arrest before a white officer shot him seven times. The shooting left Blake paralyzed from the waist down. Rioting ensued and only got worse as national media outlets descended on the scene. A state of emergency was declared. Buildings were engulfed in flames. When all was said and done, the Kenosha area had suffered $50 million in damage.

"What you are seeing now, these images, came and come in stark contrast to what we saw over the course of the daytime hours in Kenosha and into the early evening, which were largely peaceful demonstrations in the face of law enforcement," CNN reporter Omar Jimenez told viewers on August 27.

As Jimenez continued his report, a CNN chyron on the bottom

of the screen screamed, "FIERY BUT MOSTLY PEACEFUL PROTESTS AFTER POLICE SHOOTING."

But behind Jimenez was something that looked like the final scene at Nakatomi Plaza in the original *Die Hard*: cars and buildings inundated in fire to the point the night sky was almost completely glowing in bright orange. This was beyond whatever beyond parity is.

CNN wasn't alone. Here's MSNBC's Ali Velshi, likely in need of a chiropractor after bending over backward to defend the riots in Minneapolis in June 2020.

"I want to be clear on how I characterize this. This is mostly a protest. It is not, generally speaking, unruly but fires have been started and this crowd is relishing that," Velshi reported as fires raged behind him.

Mostly peaceful riots. It's like saying that serial killers are "mostly peaceful." Just think of all the potential victims they *don't murder*, after all.

The Kyle Rittenhouse trial and his exoneration were also an example of bias in broad daylight. Even before the jury in his murder trial returned a verdict, Rittenhouse was guilty, according to Representative Hakeem Jeffries (D-NY). "Lock up Kyle Rittenhouse and throw away the key," he wrote on November 10, 2021.

"The judge presiding over the trial was rooting for Rittenhouse," said CNN's David Axelrod before the verdict was rendered. "This kid has the great good fortune of a de facto defense attorney on the bench."

"Maybe the judge won't be racist during the jury instructions," wrote *USA Today* sports columnist Mike Freeman.

MSNBC got into the act, for all the wrong reasons. On one evening after the trial adjourned for the day, a producer working for the network got pulled over for running a red light after attempting to follow the jury van back to its hotel. The producer told police he was on orders from his bosses in New York to follow the van. Why? Was this jury intimidation? Or an attempt to dox a member or members in an attempt to impact the trial? The judge wisely banned the network from the courtroom for the remainder of the trial.

As for the teenage defendant Rittenhouse, he was portrayed to be a white supremacist, per then–presidential candidate Joe Biden: "There's no other way to put it: the President of the United States refused to disavow white supremacists on the debate stage last night," Biden tweeted with a video montage that included Rittenhouse.

When the verdict came down on November 19, 2021, anyone watching the trial objectively knew that Rittenhouse would likely be found not guilty. Video showed he clearly acted in self-defense after he was attacked. Yet more than a few lawmakers and media members were genuinely shocked afterward.

"Kyle Rittenhouse is living proof that white tears can still forestall justice. A murderer is once again walking free today—our system is terribly broken," wrote Representative Adriano Espaillat (D-NY).

"Fuck this murderer," Keith Olbermann, once MSNBC's highest-rated host, declared.

"#KyleRittenhouse found not guilty on all counts. This is how systems conspire to entrench #WhiteSupremacy," opined Black Lives Matter on Twitter.

Meanwhile, the cofounder of Black Lives Matter, Patrisse Khan-Cullors, suddenly had the kind of money that allows one to go on a home-buying spree, purchasing four properties for $3.2 million in the U.S. alone. This use of money creates an impression of wrongdoing. For her part, Cullors denies any wrongdoing. One of those seven-figure homes sits in a majority-white neighborhood. In 2022, it was revealed that the group had spent another $6 million for another mansion near the Pacific while using donations to pay for it in cash. In a related story, BLM raised more than $90 million just in 2020. The question is, did this money go where it was supposed to?

If this is the first time you're reading about this, you're not alone. Traditional and social media ignored the story or, in Facebook's case, censored it.

"The story was removed for violating our privacy and personal information policy," Facebook explained about the story, which did nothing of the kind.

If this had been about a founder of the Tea Party, or the pro-Trump America First PAC attached to it, would broadcast or online or print media tackle this story? You know the answer.

While these mostly peaceful riots were occurring in the summer of 2020, Joe Biden was MIA in praising law enforcement. Senator Biden was, of course, once considered to be a tough-on-crime figure in the 1990s (see: the 1994 crime bill). But in 2020, he did what was politically expedient during the campaign. Take the breathless (and ultimately false) reports of Trump ordering police to use tear gas and rubber bullets on "protesters" outside the White House in June 2020.

NPR: "Peaceful Protesters Tear-Gassed to Clear Way for Trump Church Photo-Op"
New York Times: "Protesters Dispersed with Tear Gas So Trump Could Pose at Church"
ABC News: "Police Use Tear Gas, Push Back Peaceful Protesters for Trump Church Visit"

Hmmm . . . what's missing from those headlines? Perhaps the word *allegedly* in automatically connecting Trump walking to a nearby church as the reason tear gas was used?

"He tear-gassed peaceful protesters and fired rubber bullets. For a photo," Biden tweeted on June 1, 2020. "For our children, for the very soul of our country, we must defeat him."

It turns out that this, too, was complete horseshit. From CNN *one year and ten days* later:

> The US Park Police did not clear racial injustice protesters from Lafayette Park to allow for then-President Donald Trump's march to St. John's Church last June, but instead did so to allow a contractor to install a fence safely around the White House, according to a new inspector general report.

Amazing how these revelations only came out *after* Trump was back at his Mar-a-Lago residence in Florida. By the way, why did it take the inspector general to debunk this story? Isn't that the job of any of the dozens of outlets that ran with it a year earlier?

The hypocrisy of our leaders is as stunning as it is stupid. Take Nancy Pelosi in 2020, for example. The Italian American was asked about a Christopher Columbus statue that was torn down and tossed into Baltimore's Inner Harbor. Pelosi grew up in Charm City, where her father served as mayor, hence the question.

"People will do what they do," Pelosi shrugged.

Now here's Pelosi in December 2021 when asked about rising crime in San Francisco.

"It's absolutely outrageous. Obviously, it cannot continue. But the fact is that there is an attitude of lawlessness in our country that springs from I don't know where . . . and we cannot have that lawlessness become the norm."

Wow. We've gone from "people will do what they do" to "we cannot have that lawlessness become the norm." Nancy wonders where this "attitude of lawlessness" comes from, as if she just spotted a UFO. Maybe, *just maybe* this attitude has something to do with the way many Democrats stood back and even embraced riots and lawlessness throughout 2020.

Sixteen cities set homicide records in 2021. Democrats saw the mostly peaceful riots of 2020 as an opportunity to gain points with far-left voters by supporting those protests because it was an election year. Now they want to say they're tough on crime. Good luck with that.

But the part that seriously pissed me off, and I'm certain that many Americans agreed with this regardless of ideology, is when a good chunk of the media, along with some members of the medical community, made the argument that minding COVID and social distancing was no longer a necessity, at least temporarily, in the summer of 2020.

Because the *cause* of protesting against the horrible, no good, very bad racist law enforcement makes any danger from the virus moot.

From *Politico* in June 2020:

> For months, public health experts have urged Americans to take every precaution to stop the spread of Covid-19—stay at home, steer clear of friends and extended family, and absolutely avoid large gatherings. Now some of those experts are broadcasting a new message: It's time to get out of the house and join the mass protests against racism.

If we're following along correctly, going to church was a public health threat, and keeping your small business open was a public health threat, and sending your kid to school was a public health threat. But thousands gathering and chanting in close proximity to each other was fine as long as the cause and protest *were approved*.

So, if you're keeping score at home . . .

Attending sporting events outside? Superspreader event. Trump rally? Superspreader event. Sturgis biker rally in South Dakota? Superspreader event!

Black Lives Matter protest? That's fine. Biden victory celebration? That's fine. Thousands of COVID-positive migrants illegally entering the country and being released into the general population? That's fine.

In the end, words matter, especially when discussing "mostly peaceful protests." Take this ABC News headline in July 2020 as the perfect example of a headline completely contradicting itself in broad daylight.

"Protesters in California Set Fire to a Courthouse, Damaged a Police Station and Assaulted Officers After a Peaceful Demonstration Intensified."

Now, if looking at this objectively, what word would you use to start that headline off?

a) Protesters
b) Rioters

I'd go B. Which would really anger *Time* magazine, which says the term *rioter* is racist.

"Legal and historical experts say that the word *riot* is a loaded one: While it aptly describes some events that have unfolded during the past two weeks, using it risks eclipsing the full picture by zooming in on a small, sensational slice of the action," the formerly respected magazine lectured somberly.

"In general, the word *riot* connotes meaningless violence—drunken sports fans, frenzied consumers, people have lost touch with reason and given into baser instincts. But it also has a racial dimension in the U.S., as a term that's long been used (by white people) to drum up the image of black people wreaking senseless chaos in cities," the "think piece" adds.

President Biden could do something about that portrayal. Invite officers from across the country to the White House for a major ceremony. Pledge to have their backs. Declare ALL LIVES MATTER, especially those who unselfishly exist to serve and protect. Because they're being targeted like never before, with the first three months of 2022 showing a 43 percent increase in the number of police officers being shot in the line of duty compared to 2021, which was a horrible year on this front. This obviously is beyond unacceptable.

Speaking of serving and protecting, that's what guys like John Brennan and James Comey were supposed to be doing for the American people without any kind of political bias. But after Donald Trump and Mike Pence took the reins from Barack Obama and Joe Biden in 2017, the former CIA director and FBI director and other top intelligence officers quickly showed their true colors.

15

THE DEATH OF ESCAPISM

When looking back on certain years in our lives, most just meld together. I couldn't tell you one thing that happened of significance (without googling) in 2014, for example. The year 2006 is also a blur. And 1995? Yeah, sorry. Nothing clicking at the moment. But 1998? That has several indelible memories.

We were just getting to know everything we ever wanted to know about Bill and Monica, but the couple that America truly was fixated on was Jack Dawson and Rose DeWitt Bukater, the ill-fated lovers unfortunate enough to hook up in the back of a car on the *Titanic* not long before hitting that fatal iceberg the size of the White House that somehow nobody saw until it was too late.

I personally went through a special kind of hell on a Friday night in early January that year. The girl I was dating at the time convinced me to grab dinner and a movie on the same night my friends were getting together to see No. 3 Tennessee versus No. 2 Nebraska in the Orange Bowl. You'll need to trust me on this: The Vols and Huskers were actually great then.

What made this game a bit more special is that it marked the final college game of Peyton Manning's career in Knoxville. The son of former Saints quarterback and human tackling dummy Archie Manning, Peyton was set to be the surefire No. 1 overall pick in the upcoming NFL Draft, and a bowl match-up against the vaunted Cornhuskers was must-see TV for sports dorks like my crew.

But said girlfriend at the time was quite convincing. She was (and I think still is) a prosecutor and could perform the real-life version of a Jedi mind trick. So when she said she was coming to Hoboken and wanted to see *Titanic* after a meal at my favorite restaurant, Softie-Conch defeated Sporty-Conch in a first-round TKO in terms of the itinerary for the night.

But upon hearing that I wasn't going to meet my crew at the iconic Texas Arizona sports bar in town, let's just say the ridicule from the boys was relentless the following day when meeting up for some NFL playoff games.

"You went to see *what* again?" Bill, a six-foot-three human house who drinks beer for breakfast, asked while already knowing the answer.

"*Titanic*," I sheepishly responded. "I gotta say, it was actually really good. DiCaprio is great and the Kate girl who I never saw before who played the female lead owned every scene. I really recommend—"

Popcorn and various suds from other members of the fantasy football table were thrown my way before I could complete the sentence.

Peyton and the Vols were blown out, 42–17, by Nebraska, by the way (Manning, along with his little brother Eli, who lived near me in Hoboken, would go on to win four Super Bowls combined). To this day, I don't regret my decision to see the movie on the big screen on a Friday night in a packed theater. Watching the flick on HBO (in standard definition) a few months later after every spoiler was revealed just wouldn't be the same experience. In the end, it didn't work out with my date after a few more dinners and movies. I would meet my wife in a Jersey Shore beach house seven years later, but that's a story for another book.

The reason you're briefly reading about Peyton and Leo and Kate Winslet in a book about Joe Biden and the media is to serve as a reminder that we once had this thing called escapism. *Titanic*, arguably director James Cameron's finest work, would go on to make $2.2 *billion* at the box office and dominate the Oscars that year, earning fourteen nominations to tie the all-time record set by 1950's *All About Eve.* Best Picture. Best Director. Eleven statues were captured overall, tying the record set by *Ben Hur* (1959).

To put into perspective just how big the entertainment industry was in 1998, *57 million people* watched the Oscars on one network, ABC, that year. Let's put that number in perspective: 34 million watched Joe Biden's inaugural address in January 2021 . . . on *seventeen networks.*

A little more perspective: Fast-forward to 2021.

Event: The Oscars. The venues: Union Station in downtown Los Angeles and the Dolby Theatre in Hollywood.

There was no movie anyone was buzzing about. No household-name stars were nominated, unless Anthony Hopkins—who had won his last Oscar three decades ago for *The Silence of the Lambs* and was now up for, and won, Best Actor for *The Father*—was still considered one.

America was apathetic about the 2021 telecast on ABC, with just under 9.9 million people tuning in, or nearly *50 million less* than the 1998 offering.

There wasn't even a host for the 2021 show, because the powers that be at the Academy of Motion Picture Arts and Sciences thought it was a great idea to not use one even when an unapologetic, unfiltered treasure like Ricky Gervais would have been just the person to lift our spirits.

"Most of you spent less time in school than [teen climate activist] Greta Thunberg. So if you win, come up, accept your award, thank your agent and your God, and fuck off," Gervais said as host at the 2020 Golden Globes.

Perfect.

Of course, no one paid him any attention. Anyone unfortunate enough to watch the rest of the show got to see actors, who pay big money for personal security, talk about law enforcement with disdain.

"Today the police will kill three people. And tomorrow the police will kill three people," actor Travon Free said at the 2021 Oscars, where Joe Biden also spoke to his core supporters in attendance (via video, of course).

"And the day after that, the police will kill three people because on average the police in America every day kill three people, which amounts to about a thousand people a year. And those people happen to disproportionately be Black people," Free added.

Gee, it's hard to see why mostly everyone stopped watching.

So, where did the audience go? Well, more than a few folks got tired of being lectured to. It's also no fun when your average activist actress or actor sounds no different from Maxine Waters or AOC. The factors are clear: The country is infinitely more divided, to the point that entertainment is comprised almost entirely of elite progressives while common cultural touchstones have been eliminated.

Social media plays a *huge* role in this. Twitter and Instagram, for example, allow for many celebrities to keep feeding their perpetual need for adulation and attention.

In 2022 and beyond, the overwhelming majority of "jokes" will be directed at Donald Trump. Their hope, despite all the warnings about the end of the republic if he runs and wins again, is that he gives them six more years of material upon announcing in 2023 and being in power until early 2029. Trump made them rich. It also gives these insecure performers a sense of moral superiority.

Political speeches are far easier to dredge up than comedic material. If Trump doesn't run, we're already seeing Ron DeSantis being presented as the guy who may be *worse than Trump*. And Hitler, of course.

Joe Biden, by a country mile, has provided these folks enough material to last a lifetime. Yet they still feel a need to protect him. There may be a benign joke here and there, but with a fraction of the quantity and none of the vitriol. If Biden doesn't run, rest assured that Pete Buttigieg or Elizabeth Warren or Gavin Newsom will be afforded the same exact courtesy.

How insufferable did award shows get after Donald Trump became President Trump? Here's some of the not-so-greatest hits. It's nauseating in its smugness, when it wasn't outright sick in its treatment of First Lady Melania Trump.

The 89th Academy Awards were held at the Dolby Theatre on February 26, 2017, one month after Donald Trump was sworn in as the nation's forty-fifth president.

"Flesh-and-blood actors are migrant workers," actor Gael García Bernal declared while presenting an award. "We travel all over the world. We construct families, we build life but we cannot be divided. As a Mexican, as a Latin American, as a migrant worker, as a human being I'm against any form of wall that wants to separate us. I'm against any form of wall that wants to separate us!"

How utterly reckless and ridiculous does this sound now? No wall. No borders. And millions have entered the country illegally under Joe Biden after he sent a verbal Evite to the world to drop everything and come on over for a permanent vacation, stretching our resources to a breaking point and in the process taking jobs away from those here legally.

It wasn't just the actors at the 2017 Oscars who felt the need to lecture, but the host as well. Gone are the days of emcees of actual talent, guys like Billy Crystal, in this capacity. Crystal focused on the nominees and the films and made us laugh hard in the process.

Instead, in 2017, we got the Pope of Late Night.

"I want to say, 'Thank you, President Trump,'" ABC's Jimmy Kimmel said in one of several jokes focused on Trump. "I mean, remember last year when it seemed like the Oscars were racist?"

Kimmel also said that night that we as a country need to "have a positive, considerate conversation, not as liberals or conservatives, as Americans."

Here's what this clown said six months later to CBS News: "I want everyone with a television to watch the show, but if they're so turned off by my opinion on health care and gun violence then . . . [raises his hand and waves good-bye]. I don't know, I probably wouldn't want to have a conversation with them anyway."

"Good riddance?" asked *CBS Sunday Morning* anchor Tracy Smith.

"Not good riddance," Kimmel responded. "But . . . riddance."

In case you forgot, Kimmel is the guy who wore blackface, not at a private party, but on national television while also doing an insulting imitation of NBA great Karl Malone, who once served on the National Rifle Association's board of directors. Kimmel made him sound like an uneducated idiot.

After some backlash when videos of Kimmel resurfaced of him doing this portrayal on Comedy Central's *The Man Show*, he initially tried to ignore the criticism but finally apologized. Sort of.

"I have long been reluctant to address this, as I knew doing so would be celebrated as a victory by those who equate apologies with weakness and cheer for leaders who use prejudice to divide us," the pious one said.

"That delay was a mistake," Kimmel wrote. "There is nothing

more important to me than your respect, and I apologize to those who were genuinely hurt or offended by the makeup I wore or the words I spoke."

Kimmel would go on to attempt to hammer Trump on a nightly basis. It was as predictable to watch as the Cowboys losing in January. But one night, the ABC host felt he had to push the envelope even further by mocking First Lady Melania Trump over her reading a book to young children during the annual Easter Egg Roll on the South Lawn of the White House.

"Trump heard 'egg roll' and promised to make the Chinese pay for it," Kimmel joked. "It was a fun day for the president. He got to eat chocolate. He even met a nice kid named Barron who he really liked."

At the time, Barron Trump was eleven years old.

Classy.

Kimmel then played a clip of Trump thanking Melania for her efforts on putting together the day.

"She worked so hard on this event," the president said.

"Not a chance she did one thing to help set that up. She didn't dye eggs. The only thing she's been working on is an escape tunnel," Kimmel responded before playing a clip of Melania reading to kids.

"[A]sk lots of questions about this and that," the first lady and mother read to kids seated in front of her.

"About DEES and DAT!" Kimmel mocked as his seals in the audience laughed along. "Hey Guillermo, you realize what this means?" Kimmel asked his sidekick. "You could be first lady of the United States."

It's fascinating that Kimmel, who has called Trump a racist hundreds of times on his show, thinks it is *LOL funny* to mock an immigrant's accent.

It should also be noted that Melania speaks *five* languages. Five. Yup. She's the dumb one, right, Jimmy?

It's safe to say that no president's wife has been treated more disrespectfully and disgracefully than Melania despite her conducting herself with grace and class. Another example of this treatment came after Melania had to undergo a kidney procedure in May 2018. Oddly, after about two weeks, some "journalists" and late-night hosts began wondering where the first lady could possibly be (besides, you know, resting and recovering).

"I wish that I didn't suspect that the prolonged, poorly explained public absence of Melania Trump could be about concealing abuse," tweeted Jamil Smith on June 3, 2018. He is somehow a "senior writer" with *Rolling Stone*. "I wish that it was a ludicrous prospect. I wish that the @POTUS wasn't a man with a history of abusing women, including those to whom he is married."

That tweet was liked by nearly twenty thousand people. It's also still up on Twitter.

Stephen Colbert decided to get into the act as well.

"We saw so much happen over the past week but one thing we did not see was First Lady Melania Trump," Colbert said on June 5, 2018. "As of the time we're taping the show right now, the first lady has not been seen in public for twenty-five days."

"Well, I'm not surprised," Colbert added. "It took that Shawshank guy years to tunnel out."

Do these guys just recycle jokes from their competition?

"If any First Lady 'disappeared,' you'd 'want to know where she is,'" cryptically tweeted CNN *Reliable Sources* host Brian Stelter on June 3 in pushing a segment on his weekly program.

Colbert and Smith and Stelter were all knowingly pushing a conspiracy theory, because Eamon Javers, CNBC's respected senior Washington correspondent, reported this on Twitter on May 30,

2018: "Not that this will deter the conspiracy theorists, but I saw the First Lady walking with her aides in the West Wing yesterday afternoon."

So Javers saw the first lady on May 29. That's *five days* before CNN did a missing-persons segment on its media show and almost a week before Smith alleged domestic abuse by a sitting president.

"I see the media is working overtime speculating where I am & what I'm doing," the first lady also tweeted one day after being seen by Javers. "Rest assured, I'm here at the @WhiteHouse w my family, feeling great, & working hard on behalf of children & the American people!"

In 2022 minus Trump, it's hard to distinguish late-night "comedy" from partisan news offerings these days, since the average late-night viewer can tune in to see Stephen Colbert and Seth Meyers or Trevor Noah or Samantha Bee make profoundly progressive arguments about abolishing the filibuster or about the Supreme Court.

Some headlines, picked from thousands, to share:

Mediaite: "Stephen Colbert Absolutely Destroys Mitch McConnell for His Reasoning Against Abolishing the Filibuster"

Deadline Hollywood: "Late Night Talks 'All Out War' of GOP Selecting Ruth Bader Ginsburg's Supreme Court Replacement"

The Week: "Trevor Noah Unpacks the 'Messy' New Brett Kavanaugh Sexual Misconduct Allegations"

Vanity Fair: "Seth Meyers Tears into G.O.P. Tax Plan: 'A Brazen Heist of the Country'"

The AV Club (yes, the AV Club): "Sam Bee Chokes Back Trump-Vomit as She Dutifully Rings Alarms over the GOP's Election-Rigging Plot"

Late night decided to blatantly take a side; they decided that being myopic about politics was a good idea. They have permanently lost half their potential audiences as a result.

We could see this coming right after Donald Trump won the presidency in 2016.

The venue: 30 Rockefeller Plaza, New York. The network: NBC. The program: *Saturday Night Live.*

The cold open, which often is the best part of the show, didn't begin with the usual guest host on November 12, 2016. Instead we saw Kate McKinnon at a piano, dressed as Hillary Clinton.

McKinnon, not surprisingly a huge liberal, had performed a perfect imitation of Hillary leading up to the 2016 election. But it was almost *too perfect.* Oftentimes McKinnon's Hillary was portrayed as exactly the candidate she was: an inauthentic power-hungry person who could not connect with ordinary people.

Said a 2015 review in the *New Yorker*, "The joke is that Hillary cannot approximate any of the basic emotional notes that are required—humility, geniality, relatability. McKinnon's Hillary is stiff and robotic, but also, almost at the very same time, full of sexuality and swagger."

"Citizens, *you will elect me*! I will be your leader!" McKinnon's Hillary says to the camera in a campaign video. Just as Tina Fey's portrayal of Sarah Palin helped define the governor in a negative light, McKinnon's of Hillary was doing the same. But to the folks at *Saturday Night Live*, it was all harmless fun. Just like 99 percent of the media, *no one* at 30 Rock believed Trump had any shot at being competitive, let alone winning.

That all changed when Trump was declared the winner of the 2016 presidential election on November 9. *Shock* is too soft a word to describe the feeling around newsrooms across the country. HuffPo's Presidential Forecast Model gave Hillary a 98.2 percent chance of winning on the morning of the election. The *New York Times* Upshot model put Hillary's chances at 85 percent.

Stunned can't even begin to describe how the performers at *SNL* felt after hearing the words "President-elect Trump." So the show *had* to do something. A national crisis was upon us with Trump, the former NBC guy who had guest-hosted on the show a year prior, heading to the Oval Office. This was *no time for comedy*.

Enter McKinnon as Hillary at the piano. Dressed in white. The lights dimmed. She proceeded to sing Leonard Cohen's "Hallelujah."

"I'm not giving up, and neither should you. And live from New York, it's Saturday night," McKinnon's Hillary said directly to the camera. Her voice was cracking. Her eyes welled up.

I remember thinking at the time, *What the fuck did I just watch?*

Stephen Colbert also had trouble absorbing Trump's victory during a live Election Night special that included comparing Trump's victory to (checks notes) the Civil War, World War II, and 9/11.

Really.

"Wow. That's a horrifying prospect," Colbert said as Trump appeared to be headed to victory. "I can't put a happy face on that, and that's my job."

"Outside of the Civil War, World War II, and including 9/11, this may be the most cataclysmic event the country's ever seen," political analyst Mark Halperin added.

"How did our politics get so poisonous?" Colbert later asked. "Maybe we overdosed. We drank too much of the poison."

Yep. Colbert is the virtuous one. He's the guy who elevates the national conversation.

Here's a conversation he had later on the special.

"Anything that you want to tell us about how you're feeling right now?" Colbert asked one of his guests, "comedian" Jena Friedman.

"I feel as if I'm about to give birth to a baby that's already dead," she replied.

You could hear a pin drop despite the live audience. It was that awkward (and on many levels, delicious) to watch this funeral disguised as a comedy show.

Here's Colbert, the guy who condemned poisonous politics, on his *Late Show* just three months into Trump's presidency, during one opening monologue in May 2017:

> Mr. President, you're not the POTUS, you're the "bloat-us." You're the glutton with the button. You're a regular "Gorge Washington." You're the "presi-dunce," but you're turning into a real "prick-tator." Sir, you attract more skinheads than free Rogaine. You have more people marching against you than cancer. You talk like a sign-language gorilla that got hit in the head. In fact, the only thing your mouth is good for is being Vladimir Putin's cock-holster.

Naah. Nothing poisonous and completely unhinged there. It's also patently homophobic. When the host actually received some blowback from the left, Colbert smugly doubled down. "I don't regret that. [Trump], I believe, can take care of himself. I have jokes; he has the launch codes. So it's a fair fight," he said.

Colbert would go on to host the Emmy Awards later that year.

We're a long way from Johnny Carson, who was smart enough to say that he largely avoided diving into divisive politics because it would "hurt me as an entertainer, which is what I am."

The current crop rejects that. They aren't entertainers, they're activists. Kimmel was at least honest when he wrote off half the country as a potential audience. To the elites in New York and Hollywood, they really believe that half is too stupid or too racist or a combination of both and therefore *beneath them*. Entertainers like podcaster Joe Rogan and Fox's Greg Gutfeld have benefited greatly as a result. Rogan draws more than 11 million listeners per day. Gutfeld torches Kimmel and Fallon while sometimes either beating Colbert or running about even despite having a fraction of the resources that go into late-night comedy. Why? Because half the country is thirsty for what they provide: authenticity and political incorrectness instead of piety and predictability.

Something else that has become predictable is the way CNN parrots Democrat talking points. For CNN, it's "We report. We decide."

16

THE DECLINE OF REAL JOURNALISM

The metamorphosis of CNN from a respectable news network to a de facto political super PAC for the Democratic Party began while Joe Biden was vice president. But it didn't decidedly go in that direction until Donald Trump came along. Just months after Biden took office, the whole network began to crumble from the top down. CNN put all of its chips into the middle of the table to take down President Trump. While blinded in that pursuit, they helped give us the most incompetent president of our lifetimes.

It all began forty-two years ago, when Joe Biden was just thirty-seven years old. Ted Turner's 24/7 news network, the first of its kind, would appear on something called cable TV.

Few believed it would succeed.

Over its first decade, CNN was largely chugging along, but it wasn't seen as a game changer or a true competitor to the big broadcast news entities and their big-time anchors based in New York, in the form of CBS (Dan Rather), NBC (Tom Brokaw), and ABC (Peter Jennings). That all changed in January 1991 when war broke out between the United States and Iraq over the latter's invasion of Kuwait in August 1990. CNN—thanks to a small but unrelenting team led by producers Robert Weiner and Ingrid Formanek and journalists Bernard Shaw and Peter Arnett—was the only news organization that was able to broadcast from Baghdad as the U.S. onslaught began, all courtesy of Weiner's ability to convince a difficult Iraqi government to grant him what was essentially a secure phone line out of the city to broadcast on.

Shaw, an ex-marine, along with Arnett, who had reported from dozens of hotspots around the world by that point, were there along with reporter John Holliman to provide a live audio blow-by-blow from behind a desk in the Al-Rashid hotel, as if they were Frank Gifford, Don Meredith, and Howard Cosell calling a *Monday Night Football* game.

"This is Bernie Shaw. Something is happening outside. . . . Peter Arnett, join me here. Let's describe to our viewers what we're seeing. . . . The skies over Baghdad have been illuminated. . . . We're seeing bright flashes going off all over the sky."

The glowing headlines that followed meant CNN officially became a major player along with CBS, NBC, and ABC. And off it went.

By the way, if you believe that some journalists were always as biased as many are today, check out these two paragraphs from a *Washington Post* story by Howard Kurtz on CNN's coverage from January 22, 1991. This description from an interview with Shaw seems almost foreign today:

> Somewhere in the private recesses of Bernie Shaw's mind, there may be a sense of triumph at trouncing the competition, at beating Dan Rather and Tom Brokaw and Peter Jennings on such a huge story, at Brokaw's being forced to interview him on NBC because the Big Three networks had lost their phone lines. But Shaw's puffy eyes do not betray it. "After you start reading your reviews, you run the risk of believing you're important," he says. "That can create distortions in your perceptions, in your honesty. It can color your judgment. I'm a severe critic of myself."

"You run the risk of believing you're important" is worth repeating. That kind of Journalism 101 sentiment flies in the face of CNN's own #MeToo movement during the Trump era, which is short for #LOOKATMEToo. Your chairman of the board of the #LOOKATMEtoo movement currently works for CNN. (I mean, of course, Jim Acosta, diva queen of network news.)

Up until 2002, CNN remained number one in the cable news race. But competition that came along that hadn't existed before ended its dominance, primarily in the form of Fox News and to a lesser extent, MSNBC, which both launched in 1996. Fox would take the top spot in cable news in 2002 and never looked back.

But despite the Nielsen ratings race, CNN continued to carry itself as a credible, facts-first network of integrity that leaned heavily on solid reporting, with a sprinkling of opinion and infotainment mixed in via programs such as *Larry King Live* and *Crossfire.*

In 2013, the network decided to hire former NBC Universal president Jeff Zucker to take the reins as ratings continued to be below-average at best. This gave Zucker—the former wunderkind executive producer of NBC's *Today* show—a mandate to move CNN away from its journalistic roots of more than three decades.

Zucker made his impact felt almost immediately: The weeks-long wall-to-wall coverage of the missing Malaysian plane in early 2014 was a leading example of Zucker drastically altering CNN's approach, which, in retrospect, allowed for infinitely more opinion and speculation and sensationalism prioritized over, you know, *actual journalism.*

Was Flight 370 a huge story? Of course. There were 227 passengers on board (mostly from China), including three Americans. But after a few days and no sign of debris in the massive Indian Ocean, CNN was forced to fill hundreds of hours with conversations like this:

"What if it was hijacking or terrorism or mechanical failure or pilot error? But what if it was something fully that we don't really understand?" anchor Don Lemon asked aviation expert Mary Schiavo during one interview after reading random tweets evoking comparisons to the ABC hit series *Lost.*

"A lot of people have been asking about that. About black holes and on and on and on," Lemon continued. "I know it's preposterous. But *is* it preposterous?"

Schiavo, a former inspector general for the U.S. Department of Transportation, thankfully shot that down with a basic understanding of what a black hole is.

"A small black hole would suck in our entire universe, so we know it's not that," she responded with a straight face.

CNN would go on to talk about the possibility of a "zombie plane" while devoting almost the entirety of every program to the missing jet in March and April of that year. It even went out and purchased a flight simulator to do segments when it wasn't holding a toy plane on the set.

Criticism of this approach came in hot and heavy, which was foreign to CNN. The network's credibility hadn't been questioned much to that point in its thirty-four-year history.

But suddenly the offering had become, well, *cheesy.*

New York magazine: "CNN's 9 Most Deplorable Malaysia Airlines Flight 370 Moments"

HuffPost: "CNN Uses Toy Plane to Analyze Missing Malaysia Flight"

Hollywood Reporter: "What CNN Sacrificed for Missing-Plane Ratings"

The Week: "The Daily Show Finally Figures Out CNN's Absurd Obsession with Missing Flight MH370"

Ratings went up for Zucker and CNN on its nonstop Flight 370 coverage, albeit still well short of the juggernaut that was and is Fox News. But given the swap between a cheap bump and its integrity, was it worth it?

The following year, in June 2015, the move to insert heavy doses of partisan snarky opinion into its news reports only accelerated when Donald Trump—a Zucker hire at NBC for the wildly successful *The Apprentice*—decided to jump into the presidential race. At first, CNN bear-hugged Trump's every move as ratings climbed once again. He was discussed and micro-analyzed perpetually.

But what's *that,* you say? Hillary Clinton or Bernie Sanders or Jeb Bush or Ted Cruz is doing an important policy speech somewhere? Screw it! Let's show empty podiums of Trump with chyrons blaring "BREAKING: Trump to speak soon" instead! Just about every Trump rally was shown in its entirety. It's funny to think about now, given CNN's current feelings on Trump and conservatives in general.

Trump's Republican challengers may have outraised the real estate mogul, but that was irrelevant. An analysis by SMG Delta, a firm that tracks television advertising, found that Trump was being outspent in this department by more than *8 to 1* by Jeb Bush. But because of an estimated $2 billion in free media attention afforded to Trump, he didn't need to buy much in terms of traditional ads. Why would he? The CNNs and the rest of broadcast and social media were making sure Trump got all the ubiquitous advertising he needed.

Consequently, his seventeen Republican challengers never had a shot. Trump, a walking sound bite that invariably prompts a reaction, blotted out the sun in terms of media coverage and clinched the nomination easily in April 2016.

At that point, Zucker and CNN began to worry. Bigly. Because while it was a ratings boon for the network to make Trump the centerpiece like some human version of Flight 370, there was growing concern that *the guy could actually beat Hillary* and become the nation's forty-fifth president. So Zucker unleashed the hounds (anchors/pundits) and went 150 percent in on being the anti-Trump network not named MSNBC, CBS, NBC, ABC, or PBS.

Trump, despite his longtime friendship with Zucker, clearly saw and heard enough. To this day, he has never appeared on CNN since August 2016. It also didn't matter. On his way to the presidency, Trump didn't need the network anymore as Election Day drew near. Chants of "C-N-N sucks!" echoed throughout his campaign rallies. Hillary may have been the primary opponent, but CNN and traditional media were a very close second. It's the oldest rule in politics: Attack an unpopular institution. And instead of going after, say, the IRS or big government, Trump chose an entity as unpopular as any among conservatives and many independents: "the corrupt, fake news media," at least as he refers to it.

We all know what happened on Election Night from there: Trump won all the right states to easily capture the Electoral College: Florida, Ohio, Michigan, Pennsylvania, and Wisconsin. But how did Trump win with the headwinds of almost an entire media against him while they all supported his Democratic opponent?

Well, it turns out that endorsements ain't what they used to be, because before the 2016 presidential election—and therefore before the Trump era in Washington—*57 of 59* major U.S. newspapers endorsed Hillary Clinton that year, according to *The Hill*. Only the *Las Vegas Review-Journal* and the *Florida Times-Union*, in Jacksonville, endorsed the Republican. So . . . what did all those endorsements ultimately get Hillary?

A concession speech and a set of steak knives.

The biggest newspaper in the country, the *New York Times*, for example, has not endorsed a Republican presidential candidate since Biden was in high school (Eisenhower, 1956), which means they actually endorsed heavyweights such as McGovern, Carter, Mondale, Dukakis, Gore, and Kerry. The *Washington Post* has never endorsed a Republican presidential candidate in its history of endorsements, *period*.

So how did so many in the press miss the grassroots groundswell under Trump heading into Election Night? Simple. They almost never left their ivory towers in New York and Washington, DC, and bothered to go to Michigan, Wisconsin, and Pennsylvania. The thought process by the elites, after all, is that a Republican *can't* win in those states, otherwise known as the Blue Wall. For her part, Hillary barely set foot in Michigan and never bothered to go to Wisconsin.

You know who did correctly predict a Trump win? Uberliberal Michael Moore, of all people. How? He talked to regular people on the ground in his home state of Michigan.

"I know a lot of people in Michigan that are planning to vote for Trump and they don't necessarily like him that much, and they don't necessarily agree with him. They're not racist or rednecks, they're actually pretty decent people," Moore noted before shockingly, absolutely nailing the key to Trump's appeal. Who'd have thought that Michael Moore, of all people, *got it*? He said:

> Corporate America hates Trump. Wall Street hates Trump. The career politicians hate Trump. The media hates Trump, after they loved him and created him, and now hate.
>
> Thank you media: the enemy of my enemy is who I'm voting for on November 8. Yes, on November 8, you Joe Blow, Steve Blow, Bob Blow, Billy Blow, all the Blows get to go and blow up the whole goddamn system because it's your right. Trump's election is going to be the biggest fuck you ever recorded in human history and it will feel good.

Of course, it was more than that—and Moore does the typical liberal thing of assuming the other side isn't motivated by any sort of ideology, just emotion. Justified resentment against a decadent ruling class may have been the underlying emotion, but people also loved plenty of Trump's stated policies. They wanted a border wall. They wanted America to pull out of foreign wars. Regardless, so many media elites were shocked, *shocked,* when Moore's prophecy came true on Election Night. And for Trump supporters, CNN became must-see TV on November 8, 2016. Not for providing the best news and information, of course, but because the on-air folks were simply an emotional wreck. We've already discussed Van Jones's ridiculously emotional response to Trump's victory.

"This was a whitelash against a changing country," he said. "It was whitelash against a Black president in part. And that's the part where the pain comes."

Yep. According to Jones, Trump didn't win because of his immigration stance. Didn't win on promising to secure the border. He didn't win on a promise to cut taxes. Or his position on international trade deals the U.S. had made. Or on his plan to shrink the size of the federal government and slash regulations and red tape. Or his list of potential Supreme Court candidates (releasing that short list was by far the best decision of the campaign made to remind conservatives on the fence of the stakes of the election).

To Jones, Trump didn't win because Hillary Clinton was a profoundly flawed candidate with serious authenticity and likability issues. Or because many people in the country, the kind of people not involved in politics, the kind who normally don't vote, were fed up with the establishment. The Swamp. The system.

When you really think about it, Trump beat Hillary for the same reasons Obama did for the nomination in 2008. Both men, polar opposites of each other in almost every way, represented *change*. Hope. Two guys seen as outsiders poised to upend a corrupt establishment whose sole purpose is not to serve the people, but to hold and accumulate power.

No matter. CNN wasn't interested in exploring why Trump won or how it had screwed up as badly as the polling firms in failing to capture the true pulse of the people in key states.

Instead, the network had a different idea: to call the 63 million people who elected Donald Trump closet racists who simply yearned to install a modern-day Kremlin-controlled Klansman into the Oval Office. To underscore MSNBC's and CNN's obsession with race at one point during the Trump presidency, the two networks said the word *racist* more than 4,100 times from July 14 to 21, 2019, according to a tally by Grabien, an online media production and news prep service.

Think about that for a moment: There are 10,080 minutes in a week. And for two news networks to say "race" or "racist" that many times in seven days means literally uttering the words *every few minutes.*

We all know the old children's story called "The Boy Who Cried Wolf." We saw it again, and again, when it came to crying racism in the Trump era. We all know what happened to the boy who cried wolf too often: People stopped listening. The charge of racism has lost its impact as a result. It used to be a big deal to go on television and call anyone, let alone a sitting president, a racist. No more. Under Trump, hearing about racists and racism became as common as using the words *the* and *and.*

While CNN loved putting the racist label on President Trump, they wouldn't dare do the same with President Biden. Yet there were plenty of opportunities to do so. For example, in 2021 Biden said Latinos in America were resisting vaccinations because "they're worried that they'll be vaccinated and deported."

Why would Biden think that all Latinos are automatically here illegally?

"In Delaware, the largest growth in population is Indian Americans moving from India. You cannot go to a 7–Eleven or a Dunkin' Donuts unless you have a slight Indian accent. I'm not joking," said Biden in 2006. He wasn't joking. That clip didn't get played back on CNN during the campaign. But if Trump had said it, cue the continuous loop.

"Poor kids are just as bright and just as talented as white kids,"

Biden said, also during the campaign in 2020. Van Jones stayed strangely silent. Because there are no white "poor kids," amirite?

Go back to 1977 and Biden sounds even nuttier: "My children are going to grow up in a jungle, the jungle being a racial jungle with tensions having built so high that it is going to explode at some point."

Hmmm . . . sounds kinda racist-y to me.

So, feeling emboldened as the resistance to Trump after Jones's "whitelash" rant was celebrated, CNN decided there would be no honeymoon period for the new president. Talk of the Russians handing Trump the White House began immediately, even before his inauguration, starting with the infamous Steele Dossier.

Ah, the dossier . . . Sounds like it comes straight out of a Tom Clancy novel. Such intrigue and international importance to it! The dossier! From former MI6 superspy Christopher Steele, who has the goods on Trump, right down to him peeing on hookers in a Moscow hotel room!

Of course, every major news organization—except for BuzzFeed and CNN—passed on publishing the dossier when it was shopped. Too many salacious facts, many of which couldn't be verified.

But BuzzFeed decided to publish it in its entirety anyway, because not having any real journalistic standards whatsoever allows for such things to happen. CNN followed suit in publishing excerpts. Just like that, the dossier was in the public domain, which provided the James Comey–led FBI the go-ahead to brief President-elect Trump on its contents. The unverified dossier also played an integral part in securing a FISA warrant of Trump campaign associates, including Carter Page.

Trump called it "phony stuff," "crap," and the work of "sick people" among his political opponents. He, of course, was mocked for having the audacity of doubting it. It turns out the dossier was simply a glorified document of dirty opposition research, which was ultimately paid for by the Clinton campaign and the Democratic National Committee. Imagine that: a U.S. presidential campaign tapping a foreign agent who turned to an adversary in Russia to dig up dirt on a political opponent in an effort to impact an election.

But since Democrats did it, there was no reckoning as the dossier got memory-holed, most notably by CNN. Axios explained in November 2021, "Outsized coverage of the unvetted document drove a media frenzy at the start of Donald Trump's presidency that helped drive a narrative of collusion between former President Trump and Russia. It also helped drive an even bigger wedge between former President Trump and the press at the very beginning of his presidency. . . . CNN and MSNBC did not respond to requests for comment about whether they planned to revisit or correct any of their coverage around the dossier."

Yep. The very networks that screamed about press freedom, about the importance of truth and transparency, won't even respond to requests for comment on how much they screwed up in making the dossier the next Watergate papers. Nor did they do so on the air in any meaningful way, because the allegation always gets a thousand times the play as the exoneration.

A 2018 Monmouth poll underscores a disturbing snapshot of the American people's perception of media. More than 3 in 4 respondents, or 77 percent, said they believe that major traditional television and newspaper media outlets report "fake news." That's up 14 points from the year before. Numbers mean little without context, so remember this: In 1976, trust and confidence in the media was at *72 percent approval*, according to Gallup. Now nearly three-quarters believe that traditional major news sources "knowingly reporting news it knows to be fake or false." Talk about a one-eighty.

The first hundred days of Trump's administration, according to a Harvard study, were full of almost nonstop Trump-bashing. The same study concluded that CNN led the way, along with Zucker's former home of NBC, in giving Trump 93 percent negative coverage in his first fifteen weeks in office.

This would continue through the 2020 election, with CNN running fixed town halls that were nothing but promotional platforms for Joe Biden. One such event not long before the election featured thirteen Democrats asking questions of the Democratic candidate, with just *three Republicans* getting the same opportunity. This was all by design, of course, and a preview of things to come. In his first year in office, Biden did three town halls. All with CNN. Funny how that works.

As these "town halls" (which were as fixed as a WWE match) were happening, CNN "anchors" were doing their very best to fuel an already chaotic situation in American cities during the summer of 2020. Protests raged in Minneapolis, Waukesha, Wisconsin, Atlanta, Portland, Seattle, Los Angeles, New York, Philadelphia, and other major and smaller cities. Billions in property damage resulted. Businesses were paralyzed or lost. Dozens of police officers and former police officers were either injured or killed.

Luckily for these "protesters," one CNN anchor in particular made sure to have their back, all while greasing the skids for his brother, who happened to be the Democratic governor of New York.

17

KEEPING UP WITH THE CUOMOS

Two big factors that helped Joe Biden get elected were COVID hitting the country in March 2020 and riots across the country playing out on TV screens just months before the election. Both were deadly. Both created fear. Both were good for ratings. It gave the Democratic nominee in Joe Biden another excuse to not run on his own policies but simply to cast blame on his opponent.

CNN gave him a very big assist in that department.

In the summer of 2020, Chris Cuomo, the highest-rated "anchor" at CNN, decided to give protesters who turned to violence the thumbs-up.

"Now too many see the protests as the problem. No, the problem is what forced your fellow citizens to take to the streets: persistent and poisonous inequities and injustice," Cuomo told his audience in June 2020. "And please, show me where it says that protests are supposed to be polite and peaceful. Because I can show you that outraged citizens are what made the country what she is and led to any major milestone. To be honest, this is not a tranquil time."

He would later add that "police are the ones required to be peaceful, to deescalate, to remain calm."

Lots to unpack here. For starters, the Constitution and the First Amendment specify our right "peaceably to assemble." So . . . there's that. And what an insult to the men and women in blue to say that only "police are the ones required to be peaceful." But then again, this is the same objective journalist who supported Antifa in 2018 by comparing them to (checks notes) Allied troops at Normandy.

"Anti-fascists disrupting a large gathering of white supremacists," reads the Twitter meme shared by Cuomo that shows Allied troops storming Normandy in June 1944.

"Antifa or whomever . . . or malcontent or misguided, they are also wrong to hit, but fighting hate is right," Cuomo would say on his program in the summer of 2020. "And in a clash between hate and those who oppose it, those who oppose it are on the side of right."

So as long as the cause is just, violence against others is justified. Check. But these comments were nothing compared to Cuomo's behavior in his final year at the network before his ouster.

It was the transition from winter to spring in 2020. COVID had suddenly, shockingly shut down the country in early March. New York City was especially overrun by the virus, averaging more than eight hundred deaths *per day* by mid-April. Tests were almost impossible to get. Unless you were a connected Cuomo.

We didn't know it at the time, but reporting by the *Washington Post* and others later revealed that Cuomo's older brother, Governor Andrew Cuomo, had arranged for his siblings and others close to the family to have special access to coronavirus tests administered by the state very early on in the pandemic, even going so far as to send state health officials to their homes, including Chris Cuomo's in the Hamptons, on the eastern end of Long Island.

Chris would go on to test positive for COVID in late March but continued his 9 P.M. show on CNN while suffering from the virus.

"I'm telling you I was hallucinating," he told viewers that night. "My dad was talking to me. I was seeing people from college, people I haven't seen in forever. It was freaky what I lived through last night, and it may happen again tonight."

The anchor-on-the-air-while-having-COVID angle made national headlines outside of CNN. Ratings went up as a result. But then a strange thing happened on Easter Sunday in April of that year. Chris was spotted outside a Hamptons home under construction about thirty minutes from his residence in the area at the time. Two days before, on the air, the anchor had complained about still being highly symptomatic, with low-grade fever and chills. In other words, he still could potentially shed the virus onto other people at a time when no vaccine or therapeutics were available.

The man who spotted Cuomo out of quarantine was a Long Island bicyclist, a sixty-five-year-old man who only identified himself as David, who told the *New York Post* about a verbal altercation after he saw Cuomo outside.

"I just looked and said, 'Is that Chris Cuomo? Isn't he supposed to be quarantined?'" David told the paper. "I said to him, 'Your brother is the coronavirus czar, and you're not even following his rules—unnecessary travel.'"

"'Who the hell are you?! I can do what I want!' He just ranted, screaming, 'I'll find out who you are!'" an unhinged Cuomo responded, according to David.

"He said, 'This is not the end of this. You'll deal with this later. We will meet again.' If that's not a threat, I don't know what is," David added.

Cuomo confirmed the incident on his SiriusXM satellite radio program, *Let's Get After It*, twenty-four hours later in making an argument around that because he's a famous person, that doesn't allow him to clap back at regular people talking "bullshit" at him.

"I don't want some jackass, loser, fat-tire biker being able to pull over and get in my space and talk bullshit to me, I don't want to hear it," he added.

So, yes, he admitted to breaking his own quarantine. On the air.

But if you thought an apology was coming, think again. Instead, and this is easily the most cringeworthy, dishonest, narcissistic thing you'll ever see on cable news or any station for that matter, Chris actually ended his CNN program on April 21, 2020, by showing a tape of himself "emerging" from his basement that day, almost like watching Jesus rising from the dead.

"All right, here it is—the official reentry from the basement," Cuomo told his audience while coming up the stairs in what was *most definitely* not the first time in weeks.

"Cleared by CDC," he continued (because the CDC apparently clears *individuals* during a national health crisis). Then he proceeded to wipe his brow with an open palm as he casually announced, "A little sweaty—just worked out. It happens."

So, if you're following along, Chris *decided to work out first* before coming upstairs from a basement he was allegedly trapped in for an extended period of time. You can't make this stuff up.

"This is the dream, just to be back up here doing normal things," he added.

CNN allowed all of this to happen while even heavily promoting it. Why? Because they could. Most of the media, outside of right-leaning outlets, ignored the fact that Cuomo had broken his own quarantine to the point of having a police report filed against him.

USA Today: "'I'm Back': CNN's Chris Cuomo Re-emerges from Basement After Three Weeks of Coronavirus Isolation"

People: "Chris Cuomo Emerges from Basement & Reveals Wife Is Also Out of Quarantine After COVID-19 Diagnosis"

CNN (which pretends to cover its own objectively): "Chris Cuomo returns from his basement quarantine . . . 'a day he has been dreaming of for weeks'"

Not one mention of the broken quarantine. Not one reprimand from the network. From there, Andrew would join *Cuomo Prime Time* for eleven extensive interviews that served as just about the best PR any politician could ask for. It was also a chance to portray Andrew as a true leader of the country on COVID while making Trump out to be an incompetent buffoon on the matter. "Let me ask you something. With all of this adulation that you're getting for doing your job, are you thinking about running for president? Tell the audience," Chris asked Andrew on March 30, 2020.

"No. No," the governor replied.

"How can you know what you might think about at some point right now?" Chris followed.

Remember, this is well into the primary season, with Biden on his way to the nomination. "Because I know what I might think about and what I won't think about. I won't think about it. But you're a great interviewer by the way," Andrew replied.

"Appreciate it. Learned from the best," Chris said.

"At one point, Chris, it comes down to a simple concept. It's about leadership. The experience, the wisdom, the capacity to do the job. Not just think about it, not just talk about it, not just tweet about it—do the job. And Joe Biden is a guy who does the job for real Americans," Andrew later added.

When the governor wasn't there, Chris played the attack dog on a nightly basis against Trump.

"Our money, our government, our power. We gave it to you. We don't serve you. You serve us. We have the most cases in the world, why? Because we have a big population, there is lots of density in place, and we have major foreign travel hubs. But also because we've done the least to stop the spread. And in large part, that's on Trump. He slept on this. He lied about it. And now he is not doing enough," Chris ranted on CNN in March 2020 as Trump had just sent the massive hospital ship, the USS *Comfort*, to New York (faster than anyone expected).

Regardless of the obvious conflict of interest, *Keeping Up with the Cuomos* became a big hit, with the duo even morphing into the Sammy Davis Jr. and Dean Martin on what's supposed to be a news network.

"Is it true that this was the swab that the nurse was actually using on you?" asked Chris, holding up what is now an infamous giant Q-tip as a prop. "And that at first it went into your nose and disappeared so that in scale this was the actual swab that was being used to fit up that double-barreled shotgun that you have mounted on the front of your pretty face?"

"I thought I did so well on that nasal test standing up there, she did the swab, I did not flinch. I was a cool dude in a loose mood, didn't move," Andrew replied.

This embarrassment would go on for weeks, even as reports began to emerge of a possible New York nursing home cover-up around the number of deaths that were actually occurring in those facilities. But at this point, ego and arrogance had completely taken over. This was the Cuomos' time now.

"I hope you are able to appreciate what you did in your state and what it means for the rest of the country now and what it will always mean to those who love and care about you the most," Chris said to Andrew in May 2020, while still completely ignoring the nursing home story.

"I'm wowed by what you did," Chris said, "and more importantly, I'm wowed by how you did it. This was very hard. I know it's not over. But obviously I love you as a brother, obviously I'll never be objective, obviously I think you're the best politician in the country. But I hope you feel good about what you did for your people because I know they appreciate it."

That, my friends, is unfiltered activism. No anchor like Bernard Shaw would ever make such a statement. Jeff Zucker allowed it all to happen while a majority of the media simply looked the other way. Because if Democrats like Cuomo and Biden look like the responsible ones on COVID, that's bad for Trump and the Republicans. And that was the goal.

While this was happening, Andrew Cuomo was negotiating a book deal that would eventually include an advance exceeding a reported $5.1 million. There was also talk of him being a contender for an Emmy Award (which he eventually won), all thanks to his PowerPoint presentations used for COVID-19 updates that were carried in their entirety by national media outlets.

"Your governor in New York's done one hell of a job. I think he's the gold standard," candidate Joe Biden told NBC *Tonight Show* host Jimmy Fallon at the time. "Trump waited too long to take it seriously. We're in tough shape right now."

In October, the tide began to turn. The governor released his book on COVID leadership as COVID was resurging in what turned out to be an optics nightmare.

Then came the Emmy Award that Cuomo gladly accepted, which came complete with celebrities fawning over the governor. "Governor Andrew Cuomo, you are *the man*," declared actress Rosie Perez.

"In the midst of this storm, Andrew Cuomo became the nation's governor. People across the country tuned into his press conferences every day," said singer Billy Joel.

In February, things got much worse for the governor after his top aide, Melissa DeRosa, admitted to what amounted to a cover-up around nursing home death tolls being undercounted by at least thirteen thousand in comments caught on tape.

"We weren't sure if what we were going to give to the [Trump]

Department of Justice [in terms of nursing home data] . . . was going to be used against us, while we weren't sure if there was going to be an investigation," she said.

CNN and Cuomo conveniently announced shortly thereafter that it was impossible for Chris to be a journalist on the network and cover his brother objectively.

"Let me say something that I'm sure is very obvious to you who watch my show," Cuomo said. "You're straight with me, I'll be straight with you. Obviously, I'm aware of what's going on with my brother. And obviously, I cannot cover it because he is my brother."

As for CNN itself, only when things went south for the Luv Guv did the network realize this whole brother-interviewing-brother thing might not look very good if basic integrity is your thing.

"Chris has not been involved in CNN's extensive coverage of the allegations against Governor Cuomo—on-air or behind the scenes," the network said in the spring of 2021. "In part because, as he has said on his show, he could never be objective. But also because he often serves as a sounding board for his brother."

Yeah, no shit.

It's easy to follow the money on this sordid tale at this point. Chris Cuomo got his ratings and, more important to him, adulation from his media counterparts. This obviously all benefited CNN, which cheered the Cuomo-Cuomo partnership despite the obvious journalistic ethical breach because it profited from it. Andrew Cuomo, with all the great press coverage and high approval ratings, got his Emmy and a book deal for millions up front. Joe Biden got the Democrat contrast he needed in Andrew Cuomo versus Trump.

Later in 2021, it took multiple allegations of sexual harassment and misconduct against the governor for the press to finally make Andrew Cuomo face the intense scrutiny he deserved all along regarding the nursing home scandal. One female staffer's accusations became six women, then eleven, with many of them *decades* younger than the governor.

Cuomo's approval ratings plummeted. But it wasn't until New York attorney general Letitia James's scathing, meticulous report about sexual harassment that the final coffin nail was hammered.

With Andrew expendable after having served in his role of helping to oust Trump, there was no reason for anyone to stick their necks out to protect him. The state's top Democrats, including Senators Chuck Schumer and Kirsten Gillibrand, called for Cuomo's resignation; then–New York City mayor Bill de Blasio (D) called for the governor to be criminally charged with sexual misconduct. President Biden, now over the finish line, also called on the governor to resign following the attorney general's report.

Once the *New York Times* editorial board called for his resignation, you knew it was only a matter of when, not if.

As for Chris Cuomo, he was first suspended by CNN on December 1, 2021, after it came to light via the investigation that he had advised the governor on how to discredit those accusing the governor of sexual harassment. That suspension eventually became an outright firing on December 4, after another factor emerged in the investigation: sexual harassment allegations against Chris himself, who steadfastly denies these allegations, dating back to his days at ABC News.

Game over.

It was once the hottest reality show on cable news: *Keeping Up with the Cuomos*. They laughed. They cried. They were on top of a dreary, pandemic world.

Yet, one summer later, Andrew Cuomo was out as New York's governor; his younger brother followed the same path to a permanent vacation a few months after that.

But somehow, this wasn't the whole story. Not even close. Chris was *not* happy about being fired by Jeff Zucker. If he was going down, he was going to take the network president with him. Audiences were already fleeing, making the network there to serve the Biden administration like a tree falling in an increasingly empty forest.

18

CNN'S SECRET PLOT TO SUPPORT BIDEN

In February 2022, more news emerged from CNN, which had made itself the story again. It was shocking. Jeff Zucker, CNN's network president since 2013, was out. The reason? Well, according to the network, he submitted his resignation after not disclosing a consensual relationship with a female coworker, senior executive Allison Gollust.

I remember getting ready to go on the air with the great Harris Faulkner when the news broke, about one minute before I was supposed to go on to talk about another topic. My phone was blowing up, including from some old contacts at CNN, and I quickly grabbed the statement from Brian Stelter regarding Zucker's resignation.

"Joe, you're live in thirty [seconds]," a producer said in my ear.

"As part of the investigation into Chris Cuomo's tenure at CNN, I was asked about a consensual relationship with my closest colleague, someone I have worked with for more than 20 years. I acknowledge the relationship evolved in recent years. I was required to disclose it when it began but I didn't. I was wrong. As a result, I am resigning today," reads the statement delivered by Stelter, who seemingly serves as CNN's unofficial crisis management and PR chief in addition to being the network's media critic.

This can't just *be it*, I remember thinking to myself.

"On cam in three, two, one," control said in my ear.

I read the statement after floating the possibility that there was likely a business aspect to this story (there always is). After all, CNN had lost 77 percent of its audience in January 2022 when compared to January 2021.

Imagine that: If ten people are watching the network at your average airport gate waiting for a flight, eight of them left the gate and stopped watching one year later. That's hard to do.

The great NFL coach Bill Parcells once said, "You are what your record says you are." No ambiguity. And Zucker's record was bleeding an audience like we've never seen in cable news in twelve short months.

Much of the ratings plunge had to do with the departure of Donald Trump from the stage. The central character, the bad guy CNN created and who dominated every plotline, every scene, was gone . . . partly because CNN and most of the media did everything in its power to make way for Joe Biden. Also, thanks to COVID, Joe could hide from the press and emerge at his handlers' choosing.

No Trump on CNN would be like if there were no Tony Soprano after Season 4 of *The Sopranos* and yet the show was still trying to draw an audience in Season 5. Or for those from another era, try to imagine *Dallas* without J. R. Ewing. With all due respect to Paulie Walnuts and Cliff Barnes, with no Tony or no J.R. there is no reason for many to watch, especially, in CNN's case, with trust in the network's actual reporting so low.

In 2018, Ted Koppel, one of the great journalists of any era, predicted what would happen to CNN due to its obsession with Donald Trump. "He has been wonderful for the industry," Koppel told Stelter at a panel event at the National Press Club in 2018.

"But that means what?" Stelter defensively responded. "If ratings are up, that means what?"

"The ratings are up, it means you can't do without Donald Trump. You would be lost without Donald Trump," the news veteran said.

"Ted, you know that's not true," Stelter shot back.

"CNN's ratings would be in the toilet without Donald Trump," Koppel calmly responded, drawing laughter from the audience.

"You know that's not true. You're playing for laughs," Stelter retorted before later adding, "I reject the premise that these networks are making so much money off of Trump."

But the networks did. And newspapers. And online publications. When Trump left, the bottom dropped out at most traditional outlets. Layoffs ensued. Mergers occurred just to survive. The Biden era and the critical lens on the presidency being given a break after four years of tireless work wasn't good for clicks and ratings.

"Tell me for a moment if you will—let's get away from CNN; sensitive subject," Koppel continued, while clearly enjoying the moment. "Let's go to MSNBC. Is there a moment of the day when they're not focusing on Donald Trump, or some intimately related subject?"

Stelter would later continue the conversation on Twitter to lecture Koppel (who doesn't have an account), the guy who has captured multiple Emmys and a Lifetime Achievement Award over four decades, from afar.

> "I admire Koppel, but I think the cable news biz is a lot more complex than he makes it seem. Ratings rise, ratings fall. Stories come, stories go. It's important to cover Trump and it's important to cover what comes next."

Poor Ted. How will he recover from this?

So, was Zucker's departure simply about a consensual relationship with a coworker? That's *kind of a thing* in the broadcast media business. Hell, it's kind of a thing in *any* business. And the

Zucker-Gollust relationship, according to Katie Couric, who had worked under Zucker at two points in her career, had been going on for some time. Sources even told *Rolling Stone* that the relationship went back as far as *1996*. So, no, a consensual relationship between two now-single adults that everyone already knew about was not the reason Zucker was forced out.

A few weeks later, it was reported that Zucker and Gollust were also allegedly doing exactly what they fired Chris Cuomo for: advising a sitting Democratic governor on being a COVID master and how to make Trump look like an idiot.

"As Andrew sparred on a daily basis with then-President Trump over COVID messaging, the couple provided the governor with talking points on how to respond to the president's criticisms of the New York crisis," reads a February 2022 *Rolling Stone* report.

"They also booked the governor to appear on the network exclusively, which became a ratings boon for CNN, with Chris Cuomo doing the interviewing. Cuomo and Gollust's conduct, too, would appear to mark an ethical breach for executives acting on behalf of an impartial news outlet," it added.

If I ever considered teaching a class called "How Journalism Became Activism," those two paragraphs above from *Rolling Stone* lay out anecdotally the exact argument that should be at the top of the syllabus.

Here you have, at the height of a pandemic, two top network executives and a top anchor exploiting a crisis situation for political gain. Remember (and we all do), the country was in a *horrific* place, filled with uncertainty: Restaurants, small businesses, gyms, movies . . . completely shut down. Schools shuttered. Professional and college sports halted. No therapeutics. No vaccines. Hospitals overwhelmed. If there was ever a time for news organizations to educate and inform the public, this was it.

Instead, Zucker and his executive girlfriend allegedly believed it was the perfect time to exploit the situation for political gain and to help the network's ratings. For monetary gain. The goal: kneecap a sitting president in Trump to prop up a sitting Democratic governor and the Democratic nominee in Biden.

Under Zucker, CNN gave virtually *everyone* carte blanche, with inmates running the asylum while the ethically challenged warden simply looked the other way.

This was best illustrated by Jeffrey Ross Toobin, who truly showed that getting the shaft is a hard thing to do if you're squeezing the right hand internally. Toobin, you may recall, was seen pleasuring himself on a work Zoom call that included female coworkers in October 2020. But CNN didn't fire him. They didn't even suspend him without pay. Instead they sidelined him for a few months before bringing him back on the air and providing another kind of happy ending.

In 2022, the collapse of CNN was now complete: Viewers, mostly gone—by the end of June 2022, the network couldn't draw even 650,000 viewers on most nights. Reruns of everything from *Gunsmoke* (which debuted sixty years ago) to cartoons like *Peppa Pig* and *SpongeBob* were beating CNN, and easily. Its top-rated anchor, Chris Cuomo, gone. Its network president, gone. Its integrity in shambles after presenting so much opinion under the guise of news. Its credibility in tatters after allowing so many of its on-air people to make themselves the center of attention.

CNN announced in February 2022 that Chris Licht was selected to replace Zucker as CNN president. New management promised to move CNN away from partisan programming (which its "journalists" insisted didn't exist in the Zucker era) in favor of a return to the old CNN.

But Licht comes as a former executive producer of *The Late Show with Stephen Colbert* on CBS, which is perhaps the most partisan, pious, predictable political program on television.

Forget comedy. Colbert and Licht pushed an outright vitriolic product to its viewers almost exclusively focused on Donald Trump and Russia.

On one occasion, Colbert flashed a Nazi salute in mocking Trump as one. This kind of stuff happened nightly. Now we're told Licht is the guy who's going to remove partisan opinion from CNN? Does this mean the network and others in broadcast media finally begin to cover Democratic administrations and a Democratic Congress the same way it did the GOP ones before them?

Stay tuned. On second thought, don't. Just to show how deep the collapse has been for CNN, its streaming service, CNN+, debuted on March 29, 2022. Despite CNN spending at least $400 million on it, it collapsed in less than three Scaramuccis, or twenty-eight days.

This is CNN.

19

THE CONSERVATIVE ANTI-TRUMP MEDIA ARMY

They may be vastly outnumbered, but they are ubiquitous. At least in the world of political TV.

They are conservatives, or at least bill themselves that way, but they don't support conservative lawmakers. They probably lack principles. Many may be grifters. But they know how to speak in provocative sound bites. They are former Republican strategists, although finding anything resembling a big win for their clients is difficult. They are portrayed on broadcast news as #NeverTrumpers or "principled conservatives."

They've created quite the cottage industry for themselves: Rebels without a party. Profiles in courage. Those speaking truth to power without fear or favor.

Yeah, right. Because these folks are the phoniest of the phony on TV and particularly cable news.

The anti-Trump "Republicans." The act put on by these folks on networks like CNN and MSNBC is truly the best unintentional comedy out there. Here's the template:

"(Insert faux Republican here) is speaking out in putting country over party. Trump is a tyrant, and our next guest is bravely bucking the GOP as Americans are *outraged* that Trump (insert hyped controversy du jour here)."

The best example of this cottage industry, this niche, can be found with the Lincoln Project, the professional grifters who use the old Michael Avenatti rhetorical playbook while posing as the aforementioned "principled conservatives" on television on a regular basis.

So here's a thought: If you would back Biden over Trump, you may not be a Republican. But if you'd back Terry McAuliffe over Glenn Youngkin, there is *no way* you are a Republican. And the Lincoln Project did just that.

This is also a group who harbored cofounder John Weaver, sixty-two, who was accused by at least *twenty-two underage boys* of various indiscretions, including offering some of them employment for sex, according to multiple reports in February 2021.

Also, according to publications ranging from the Associated Press to the *Washington Blade*, Lincoln Project executives such as Twitter stars George Conway and Steve Schmidt knew all about the allegations against Weaver for months before the *Times* story broke it wide open.

From the *Washington Blade*:

> The Lincoln Project's leaders, amid the unfolding scandal of co-founder John Weaver soliciting sexual favors from young men, have asserted they were unaware of his indiscretions until [January 2021], but electronic communications obtained by the Washington Blade call that claim into question and suggest some Lincoln Project executives knew about the texts as early as last summer [2020], but took no substantive action in response.

When in doubt, always follow the money. There were simply too many millions coming into the Lincoln Project's coffers to let a little *our-cofounder-is-soliciting-teen-boys-for-sex* scandal get in the way.

But the death knell may have come from a February 2021 Associated Press investigation that found a good chunk of the money the Lincoln Project had raised to battle Trump or Republicans or conservatives had gone, well, *somewhere else* besides advertising, marketing, and overhead costs.

> Of the $90 million Lincoln Project has raised, more than $50 million has gone to firms controlled by the group's leaders. . . . Since its creation, the Lincoln Project has raised $90 million. But only about a third of the money, roughly $27 million, directly paid for advertisements that aired on broadcast and cable, or appeared online, during the 2020 campaign, according to an analysis of campaign finance disclosures and data from the ad tracking firm Kantar/CMAG.

So where exactly did the other $60 million or so go? Even if $20 million was allocated for operating expenses (and that's being generous), that still leaves $40 million or so that went somewhere else.

Undeterred by all the allegations of impropriety, the Project marched on in 2021.

"We're coming for you, @GlennYoungkin," it tweeted ever-so-ominously at the Republican gubernatorial candidate a few weeks before Election Day in Virginia.

At the time, Youngkin was down double digits to Democrat and Clinton royalty Terry McAuliffe. Then came what could be arguably the stupidest political stunt ever by any political group to attempt to influence an election (and that's saying something).

It was just a few days before the Election Day and Youngkin

clearly had seized the momentum. So the Project did what any principled conservatives would do: They sent staffers to a Youngkin rally to pose as white supremacists supporting Youngkin, right down to the torches. Fortunately, the gambit spectacularly backfired. After Youngkin was well on his way to victory on Election Night, I let my feelings be known when appearing on *Hannity* on Fox News in November 2021.

"Those—and I'll say this generously—scumbags, who concocted a white nationalist hoax against Glenn Youngkin in the most pathetic, inept political stunt we've seen in modern politics, only helped seal McAuliffe's fate. So that's very delicious. You could go in front of your home right now and burn money in the street before donating to those clowns and get the same return on investment."

I'm usually relatively measured when doing my Fox News hits, because I think (from observing this industry for a long time) that being over-the-top and emotional with every appearance diminishes the moments that actually truly deserve such passion. This was one such moment. And for me, at least in this instance, I was sick of the sophomoric vitriol and self-righteousness of those in this business who knowingly lie on the air, knowingly try to start race wars, get caught, and still get rewarded with more appearances and more book deals.

Take George Conway, for example. He cofounded the Lincoln Project along with Weaver. So when the disgusting Weaver allegations broke, someone finally decided to ask Conway about it.

"It's terrible and awful," Conway told MSNBC's *Morning Joe* when gently asked about the Weaver allegations. Then he added, "I didn't know John very well. I frankly only spoke to him a couple times on the phone early on in the Lincoln Project."

Oh please. You didn't know him very well, huh, George? You

only cofounded this cash cow with Weaver in raising more than $90 million with him in 2020 going into the presidential election. Oh, by the way, there's also an op-ed in the *New York Times*, "We Are Republicans, and We Want Trump Defeated," that *you cowrote* with that guy you *don't really know*.

Conway suffered no loss of appearances or Twitter prestige from any of this, and he continues to be a fixture on MSNBC and CNN to talk about Trump despite the forty-fifth president being out of office for quite some time now.

Here are some others outside of the Lincoln Project who also claim to be conservative but are anti-Trump and somehow *pro-Biden* and get plenty of airtime playing this ridiculous role.

"Former Republican strategist" Ana Navarro, who never actually ran any notable campaign, has become the loudest person on cable news (CNN).

Navarro cohosted a fundraiser for Biden before the 2020 election after endorsing him (because "Republicans" always do that). Before you say this was just an anti-Trump one-off, she also endorsed Democrat Andrew Gillum over Republican Ron DeSantis in the 2018 Florida gubernatorial race. Gillum lost, of course. So CNN signed him not long after. Of course.

It's moves like endorsing Democrats like Biden and Gillum that are supposed to make Republicans like Navarro look like rebels. And rebels who talk trash about conservatives get lucrative TV contracts. The list goes on in terms of alleged Republicans/conservatives supporting Democrats despite for decades having supported smaller government, law and order, strict border laws and enforcement, and school choice, which are the exact opposite tenets of the Democratic Party: Bill Kristol, Jennifer Rubin, Max Boot, George Will, Nicolle Wallace . . . It's all so phony.

This whole farce was originally drummed up by CNN and

MSNBC execs to create an illusion: that Trump was actually loathed by his own party, when in fact his Republican support was always around or above the *90 percent* mark throughout his presidency. Joe Biden dropped into the 70s in terms of his intraparty support in 2022, but we don't see too many disgruntled Dems on the air talking about how awful he is, now do we?

A good example of how the media reacts when Democrats turn against Republicans is perfectly illustrated by John McCain, Joe Manchin, and Kyrsten Sinema. In 2017, during Trump's first year in office, then-senator McCain gave that infamous thumbs-down when voting against repealing and replacing Obamacare. The bill would fail 49–51 in the Senate, marking a huge embarrassment to Trump and Mitch McConnell, who felt they had the votes going in.

On cue, the media absolutely embraced it, as discussed earlier:

Washington Post: "Analysis: The iconic thumbs-down vote that summed up John McCain's career"
CNN: "The 'thumbs down' health care vote that enraged Trump is John McCain's lasting legacy"
Politico: "McCain returns—backing and blasting his own party"

Now compare that to the media reaction from the same publications when two Democrats, Manchin and Sinema, did exactly what McCain did in defying their party on principle as it pertained to Joe Biden's prize proposal: Build Back Better, a $5.5 trillion social spending bill at a time when inflation was beginning to skyrocket.

Washington Post: "Despite Manchin and Sinema, Democrats are more united than they've been for decades"
CNN: "With help from Manchin and Sinema, a Republican revolution from below is driving national policy"

Politico: "Manchin, Sinema leave Dems in lurch as Biden agenda teeters"

Quite the contrast, ain't it?

But the hypocrisy is no longer surprising. The next time you see a "Republican" or "conservative" bashing Trump or DeSantis or anyone in the party, ask yourself this:

What is their résumé? Have they accomplished consequential achievements in the past for the party, for conservatives? What station or newspaper is this "analysis" occurring on or in? Are they just telling their audience and readers not what they *should* hear, but what they *want* to hear?

We all know the answer to that last one.

20

EVEN DEMS DON'T LIKE BIDEN

Donald Trump's polling numbers were always as polarized as they come. Liberals largely loathed him, while conservatives embraced him at record levels. But compared to Joe Biden? Trump was Mr. Popularity. Even Democrats find Biden underwhelming.

Contrast that with Republican opinions of Trump. Throughout the Trump presidency, the forty-fifth president's approval among Republicans hovered in the 90 percent range, according to Gallup. And when entering the 2020 election, despite a largely depressed country due to the coronavirus pandemic that shut down businesses and schools for six months, Trump's approval among voters identifying as Republican sat at 95 percent.

So, how about Joe Biden? This is the guy who received 12 million more votes in 2020 than his old boss, Barack Obama, did in 2008. *Surely* the Democratic Party is supporting Biden with the same vigor Trump was supported with during his time in the Oval.

Not. Even. Close.

A *New York Times* poll taken in July 2022 showed 64 percent of Democratic and Democratic-leaning voters *do not want* Biden to run in 2024.

Again, this is a *New York Times poll*, underscoring how a majority of Biden's own party doesn't want him in office too much longer.

It got worse in the summer, when a YouGov poll showed just 21 percent of voters wanted the guy who got 81 million votes to run again.

When Joe Biden does get support, *tepid* is the only word that comes to mind.

For instance, the Democratic Congressional Campaign Committee posted a graph in 2021 purporting to show a huge drop in gas prices as a result of Biden's leadership on the issue.

"Thanks, @JoeBiden" was the caption.

Problem is, anyone with eyesight and a halfway functioning brain could see the drop in price amounted to a whole *two cents.*

Then there's this *USA Today* beauty that deserves to be framed somewhere: "Sorry progressives, but Biden isn't doing that bad. Remember Trump? Politics is like a law school exam: Pick the least wrong answer."

We live in the greatest country in the world, and we've been reduced to *pick the least wrong answer*?!

But the top prize in this category goes to Jennifer Rubin, the formerly conservative columnist for the *Washington Post* who has become arguably the top cheerleader for this administration. Rubin often gets retweeted by White House chief of staff Ron Klain for making arguments like this, which are intended to prop Biden up but actually just draw attention to his tremendous weaknesses.

"Opinion by Jennifer Rubin: If It Weren't for Inflation, This President's Economic Performance Would Be Unmatched," reads the *Post*'s Twitter headline.

It's the "Other than that, how did you enjoy the play, Mrs. Lincoln?" moment of our time. Inflation impacts everyone. Most Americans blame Biden's policies for it. Yet here we have a "leading" columnist saying that if you take away that very big thing that is an unwanted tax on the entire population, Biden would be the greatest president on economic performance, like, ever.

Few prominent people on the Democratic side had the guts to say what they actually felt about Biden during the 2020 campaign.

But one interview on CNN early that year with Judy Sheindlin, otherwise known as Judge Judy, stands out now. Sheindlin made about $50 million per year through syndication as Judge Judy and appears to be a pragmatic person who represents more of a Kennedy or Bill Clinton Democrat of yesteryear.

"This is [Biden's] third run for president. I think he's made a lot of friends. I think people feel he's sort of safe. But my question is, if you had to have a heart valve replaced, would you want a nice guy to do it? Or do you want the best? I don't see that a nice guy should be the president of the United States. I think that a great guy, by what he's done, should be the president of the United States."

Hmmm . . . demanding greatness. Seems like a prerequisite we can all agree on when choosing a president. Instead, most on the left settled for *the least wrong answer*.

So, what is the difference between Trump and Biden? The big difference is that Republicans wanted Trump, but Democrats tolerated Biden. He was the safe option. Mr. Normal. Now, in reality Joe is so far left that this impression isn't accurate at all, but as we've argued throughout, he's a good liar. He knows how to spin a tale. So when good old Uncle Joe's radical liberal policies end in disaster, as they always do, Democrats blame his normalcy instead of his ridiculous ideology. Given the people they're considering for 2024, they're only going to make that problem worse.

When looking at the bench on the Democratic side, it's hard to get too jacked up as far as alternatives are concerned.

Stacey Abrams.

Stacey Abrams is a politician arguably more beloved by traditional media than any other, much like Barack Obama was in 2004. But Obama wasn't a losing candidate then. Abrams already is. She ran for Georgia governor in 2018 and lost to Republican

Brian Kemp by 55,000 votes, only to commit what we were told after the 2020 election was the most chilling attack on democracy anyone in American politics could commit: declaring that an election was stolen.

"Concession means to acknowledge an action is right, true or proper. . . . I cannot concede," Abrams declared following her 2018 election. She also called Georgia secretary of state Brad Raffensperger "corrupt." Sound familiar?

"I have no empirical evidence that I would have achieved a higher number of votes. However, I have sufficient, and I think legally sufficient, doubt about the process to say that it was not a fair election," Abrams would say six months later in April 2019.

"I believe it was stolen from the voters. I just said it can't happen again. And that has been my mission for the last two years," Abrams repeated in November 2020.

In May 2022, a funny thing happened on the whole voter suppression front after Georgia's senate and gubernatorial primaries: Final tallies showed a 212 percent increase in voting from the presidential primary in 2020 and a 168 percent increase from the 2018 elections. Nobody is being suppressed and prevented from voting, especially in Georgia.

Without her signature issue, Abrams will likely lose her second run at the Georgia governorship if polling numbers seen over the summer hold. Biden is also in the 30s on approval in the Peach State, a state he somehow won, dragging down the Abrams ticket further. If she chalks up another "L," count her out.

Okay . . . so how about that Democratic governor, also absolutely beloved by the media in 2020, who was rumored to replace Biden at the top of the 2020 Democratic ticket: Andrew Cuomo?

The former Luv Guv . . . not available for comment after resigning in disgrace thanks to multiple sexual harassment allegations against him, along with a nursing home scandal that costed thousands of lives.

We also hear a lot about Pete Buttigieg, the former mayor of the (relatively tiny) college town of South Bend, Indiana. He's now somehow transportation secretary after being nice enough to get out of Biden's way during the primaries and endorsing him. "Somehow" applies here because Buttigieg could barely handle transportation in South Bend, as I wrote earlier in this book.

So who else do we have? Governor Gavin Newsom of California?

He had to spend major time and resources just to avoid being ousted in deep-blue California during a recall election in 2021 (Obama, Biden, Harris, and Warren all ran to the Sunshine State to save the governor from radio talk show host Larry Elder). Homelessness and crime are rampant in California. Taxes are through the roof. And Newsom, perhaps the hypocrite to end all hypocrites, had a big issue with wearing masks in public places while demanding kids stay in theirs.

Besides, what would Newsom's 2024 bumper sticker be?

NEWSOM: He'll do to America what he did to California!

In other words, scratch him off the list.

There's always good ol' Bernie, of course. But Sanders was born before Pearl Harbor was attacked. If he ran and somehow served two terms, we'd have a ninety-something-year-old president. And the whole Democratic socialist thing just doesn't seem to be working out, if Biden's Build Back Better debacle was any indication. So he's out.

The rest of the choices are like a selection of bad movies, such as Senator Elizabeth Warren, who starred in what was easily, easily the most cringeworthy campaign ad in modern history.

Warren (in her kitchen when suddenly hit with an "impromptu" thought): "Hold on a sec. I'm going to get me a beer."

A few moments later, her husband, Bruce Mann, just happens to walk in. Because, you know, *he lives there.*

Warren: Hey, my husband, Bruce, is now in here. You want a beer?
Mann: No, I'll pass on the beer for now.

Voters eventually passed on Warren as well.

So who's left? Spartacus? Senator Corey Booker is my senator here in New Jersey. He's also a cartoon diva. Tulsi Gabbard? She is easily the most sensible and authentic of any Democrat right now, so much so that she spoke at the Conservative Political Action Conference (CPAC) earlier this year. Beto O'Rourke? How many elections could one lightweight lose in an adult lifetime?

That leaves the 2016 Democratic nominee, Hillary Clinton. She's seventy-four years old, which is like being a spring chicken compared to Biden. She also said her election was stolen from her (by the Russians, of course), but that's okay because when anyone but Trump talks about stolen elections, many in the media either cheer for it or conveniently look the other way.

A notable sign that Hillary is dipping her toe in the 2024 pool came after her bizarre decision to read her 2016 victory speech for something called "MasterClass." It was one of the most cringe-worthy things you'll ever see outside of an Elizabeth Warren campaign ad.

Here we have a former New York senator, secretary of state, and Democratic presidential nominee reading a speech for an election she lost. One would think that eventually Hillary, more than five years later, would show some semblance of integrity. Some humility. Some maturity. And not talk about her loss so often anymore.

Instead, here she is, a losing candidate reading an old victory speech for a victory that never happened. In case you're asking if any losing candidate had done anything like this before, the answer is likely no.

Since the election, Hillary has blamed her loss on misogyny, sexism, voter ID laws, Bernie Sanders, former FBI director James Comey, and Matt Lauer (yes, *Matt Lauer*), along with dozens of other factors. She hasn't blamed neglecting to campaign in Wisconsin, not having any clear campaign message, or deciding "I'm with Her" was a grabby campaign slogan, but whatever.

It's a five-year public therapy session in broad daylight. In a sane world, she would have been laughed off the stage for reading such a speech. But this felt more like a trial balloon in an effort to see if there is still an appetite for the Clinton brand.

Hillary always seemed to believe the mantle of "First Female President" was her birthright. Given how pathetic the field is on the Democratic side with or without Joe Biden, she may just get a second chance at winning the office her husband so famously made infamous.

But things could still turn around for Biden. After all, this is the guy who promised not only to control COVID, but to cure cancer.

No, really . . .

21

FAIRNESS IS OVERRATED

One of the biggest problems our country faces is a media class that's completely blind to its own bias. A 2021 Poynter study of 167 journalists found that they all thought "journalism's first obligation is to the truth" but were split on whether they thought "objectivity" was also important to journalism.

The study's authors clearly wanted us to think that objective (what they called "traditional") journalists and "activist-minded" journalists were basically the same thing. "Journalists who want to express their political views on social media, engage in activism and ally themselves with social justice protesters value truth as much as journalists who seek to maintain a neutral, dispassionate approach to the profession," read the summary, without observing that the two classes seemed to have a *very* different idea of what "truth" was.

That said, even in opinion journalism, the censorship has grown harsh.

Sometimes the best way to make a big point is to use a simple example. On the topic of media making the metamorphosis from journalism to activism, here is one.

Lester Holt, the anchor for the *NBC Nightly News*, can say something like this while accepting an award and actually get applause for saying it.

"I think it's become clear that fairness is overrated. . . . The idea that we should always give two sides equal weight and merit does not reflect the world we find ourselves in," Holt said in 2021, not long after Biden was inaugurated.

Fairness is overrated.

"The idea" of giving both sides equal weight.

Hey, give the guy credit. He's at least being honest about how many in newsrooms see journalism today. When stories emerged of Black Lives Matter using millions upon millions in donations to purchase opulent mansions in California, suddenly the outspoken Holt went mute. The *NBC Nightly News* didn't touch those stories. Because, I guess, fairness is overrated. Hawk Newsome, the head of Black Lives Matter Greater New York, even called for an investigation into the organization's finances and fundraising, making the stories even more newsworthy. In particular, he was worried about the organization's cofounder, Patrisse Khan-Cullors.

"If you go around calling yourself a socialist, you have to ask how much of her own personal money is going to charitable causes," he told the Associated Press. "It's really sad because it makes people doubt the validity of the movement and overlook the fact that it's the people that carry this movement."

But Holt doesn't broach any of these points. Because why bother with both sides, right?

Here's another example. Republican senator Tom Cotton of Oklahoma is a measured guy and has served this country with honor. Grew up on a cattle farm in a rural area. Straight-A student. Played basketball on his high school basketball team. Attended Harvard undergrad, where he served on the editorial board of the school newspaper, the *Harvard Crimson*. Graduated in three years, magna cum laude.

Cotton went on to attend Harvard Law School before joining

the army. Served in Iraq and Afghanistan. Awarded the Bronze Star. Ran for Congress at age thirty-five. Won. Ran for the Senate two years later and defeated a two-term Democrat, Mark Pryor. Married that same year; currently has two kids. In 2019 he wrote a book, *Sacred Duty: A Soldier's Tour at Arlington National Cemetery*.

When looking at résumés, it's hard to find one more complete than Cotton's.

In short, he's not an idiot. He may very well be president one day.

So when Cotton went to the *New York Times* in 2020, amid rioting and looting and chaos in American cities over the police killing of George Floyd, with an op-ed titled "Send In the Troops: The nation must restore order. The military stands ready," it was coming from someone immensely qualified to weigh in. In the piece, Cotton called for a more aggressive response as riots spun out of control, including considering sending in the military, to quell the instances of violence and looting seen amid the largely peaceful protests after the death of Floyd under policy custody.

Cotton in the *Times*, June 3, 2020:

> Outnumbered police officers, encumbered by feckless politicians, bore the brunt of the violence. In New York State, rioters ran over officers with cars on at least three occasions. In Las Vegas, an officer is in "grave" condition after being shot in the head by a rioter. In St. Louis, four police officers were shot as they attempted to disperse a mob throwing bricks and dumping gasoline; in a separate incident, a 77-year-old retired police captain was shot to death as he tried to stop looters from ransacking a pawnshop. This is "somebody's granddaddy," a bystander screamed at the scene.

More Cotton:

> According to a recent poll, 58 percent of registered voters, including nearly half of Democrats and 37 percent of African-Americans, would support cities' calling in the military to "address protests and demonstrations" that are in "response to the death of George Floyd." That opinion may not appear often in chic salons, but widespread support for it is fact nonetheless.

Logical stuff. You can agree or disagree with it. That's one of the primary points of an op-ed like this: to get people talking. Thinking. Perhaps providing a different and unique point of view. To spur debate.

But remember, this is the *New York Times*. And its newsroom is like an All-Star team filled with the wokest of the woke. On cue, several *Times* journalists went to social media to express their outrage with the paper's decision to publish Cotton's piece, including 1619 Project creator Nikole Hannah-Jones, advice columnist Roxane Gay, and opinion page staffer Charlie Warzel.

You may recognize Hannah-Jones the most. She's the "journalist" and "historian" who wrote in the *New York Times Magazine*'s 1619 Project that the Revolutionary War was fought over (checks notes) *preserving slavery* and that our history as a republic is based on white supremacy. The project would go on to (checks notes again) win a Pulitzer Prize for Commentary, despite many respected historians completely disagreeing with her premise.

"As a black woman, as a journalist, as an American, I am deeply ashamed that we ran this," Hannah-Jones tweeted regarding the Cotton piece.

"i feel compelled to say that i disagree with every word in that Tom

Cotton op-ed and it does not reflect my values," Warzel wrote before linking to his own piece accusing police of suppressing free speech.

The editorial page editor of the piece, a news veteran named James Bennet, said that he personally opposed sending in troops and that "my own view may be wrong" on publishing Cotton's opinion piece, but he defended the decision to publish it.

"It would undermine the integrity and independence of the *New York Times* if we only published views that editors like me agreed with, and it would betray what I think of as our fundamental purpose—not to tell you what to think, but to help you think for yourself," he wrote in his defense.

Cotton originally praised Bennet and the *Times* for not backing down because those on the far left took issue with his opinion.

"I will commend the *New York Times* leadership," Cotton told Fox News at the time. "We obviously don't agree on very much. But, in this case, they ran my opinion piece with which they disagreed. And they've stood up to the woke progressive mob in their own newsroom. So, I commend them for that."

All's well that ends well, right?

Nope.

Bennet, also a Pulitzer winner, was no longer with the paper two days later. He first joined the paper in 1991, nearly *three decades ago*, and just like that, he was out. The *Times* also threw a junior editor, Adam Rubenstein, under the bus, naming him as responsible for the piece, which they said "did not meet standards," in an article reporting on the internal kerfuffle. The twenty-five-year-old Rubenstein left the paper six months later, and who can blame him when it was clear his colleagues were so eager to stab him in the back?

For publishing an op-ed that a majority of the country *agreed* with. According to an ABC News/*Washington Post* poll taken just before Cotton's piece was published, 52 percent of Americans supported deploying the military to control violent protests.

In the end, the only poll that mattered at the *Times* belonged to activists in the newsroom posing as journalists.

Internal woke mob 1, veteran award-winning editor 0 (Final).

Here's what the *Times* added to the top of Cotton's op-ed. It's comical coming from what is considered by most journalism schools as the top of the journalism food chain.

> After publication, this essay met strong criticism from many readers (and many Times colleagues), prompting editors to review the piece and the editing process. Based on that review, we have concluded that the essay fell short of our standards and should not have been published.
>
> Beyond those factual questions, the tone of the essay in places is needlessly harsh and falls short of the thoughtful approach that advances useful debate. Editors should have offered suggestions to address those problems. The headline—which was written by The Times, not Senator Cotton—was incendiary and should not have been used.

Ah, the tone was *needlessly harsh.* But the *Times* had no issue with these beauties in its opinion section:

> April 13, 2018: "Tethered to a Raging Buffoon Called Trump"

The author, Roger Cohen, predictably compares Donald Trump to Adolf Hitler (which isn't needlessly harsh or anything).

July 30, 2020: "Fascism: A Concern"

The illustration below the headline features Brazilian president Jair Bolsonaro, Adolf Hitler, and Donald Trump (which isn't needlessly harsh or anything).

And hey, we can't leave out a comparison to another mass murder of millions, now can we?

September 24, 2020: "Trump's Stalinist Approach to Science"

The author, Paul Krugman, accuses Trump of intentionally killing his own citizens (which isn't needlessly harsh or anything).

There are dozens upon dozens of other examples, but the point is clear: The *New York Times* had just officially created a safe space for the far left by bowing to the likes of Hannah-Jones and others in the newsroom.

Oh, and here's the most delicious part. After the Canadian trucker protest in Ottawa made international news and was cheered by many conservatives in the U.S., the *Times* asked this question in a tweet shared by Tom Cotton:

> As protests stretch on in Canada and truckers block supply chains with the U.S., some Canadians are asking: Why hasn't Prime Minister Justin Trudeau ordered the authorities to quash the demonstrations?

Great question, guys! Maybe Cotton can write an op-ed on this sometime.

But just like CNN, these people will insist that they are objective.

Down the middle. And only exist to hold truth to power. We don't root for a side, no sir!

In its endorsement of Biden, here's what a member of the *Times* editorial board had the audacity to say without bursting out laughing.

"This election is not about Democrat or Republican," editor Mara Gay told MSNBC in 2020. "This is really about right and wrong and saving the soul of the nation."

Tom Cotton found this as amusing as you just did.

"I mean, what a joke," Cotton chuckled at the time while reacting to Gay's sound bite on *Fox & Friends*. "Do these people have any self-awareness? But if they're really not a partisan newspaper, I have got news for the *New York Times*: I will be submitting several new op-eds in the coming days that they can publish as well," he added.

You're not going to believe this, but those op-eds never got published. As for Mara Gay, here's just how out of touch she is while existing in the elite bubble that is New York City.

"I was on Long Island this weekend visiting a really dear friend, and I was really disturbed. I saw, you know, dozens and dozens of pickup trucks with explicatives [*sic*] against Joe Biden on the back of them, Trump flags, and in some cases just dozens of American flags, which is also just disturbing. . . . Essentially the message was clear. This is my country. This is not your country. I own this," Gay "reported" on MSNBC's *Morning Joe*.

"That is the real concern. Because, you know, the Trump voters who are not going to get on board with democracy, they're a minority," she later continued. "You can marginalize them, long-term. But if we don't take the threat seriously, then I think we're all in really bad shape."

"Disturbed" by Trump flags. American flags. Just like with

James Bennet, if Mara Gay had the power, she would have these people evicted from Long Island.

Her goal also is to "marginalize" Trump voters.

All 74 million of them.

Those flag wavers are truly a threat to democracy.

The paper of record, my ass.

This is how the sausage is made in media. Whether it's the lie that is the media's and Biden's "Don't Say Gay" campaign against the threat that is Ron DeSantis, or Lester Holt being the Jen Psaki for Black Lives Matter, or blatant character assassination of anyone from the right seeking office, this kind of bias in broad daylight is now the norm, not the exception.

If you see this happening, call out that anchor, that reporter, that journalist on social media. Because know this: Almost everyone in this business obsesses over what viewers and readers think of them on social media. They see everything. But don't get personal, keep it civil, and make your point based on the merits and facts. You'll be surprised how impactful this can be.

You have more power than you think.

22

OBAMA AND BIDEN'S WAR ON THE PRESS TRUMPS TRUMP

Barack Obama hadn't looked this happy since Election Night 2008.

The venue: The Washington Hilton
The event: The final White House Correspondents' Association Dinner of President Obama's time in the White House
The date: April 30, 2016

"Taking a stand on behalf of what is true does not require you shedding your objectivity. In fact, it is the essence of good journalism," Obama told the audience of reporters, news executives, cable news hosts, Hollywood celebrities, and Tony Romo. "It affirms the idea that the only way we can build consensus, the only way that we can move forward as a country, the only way we can help the world mend itself is by agreeing on a baseline of facts when it comes to the challenges that confront us all.

"So this night is a testament to all of you who have devoted your lives to that idea, who push to shine a light on the truth every single day," he continued. "So, I want to close my final White House correspondents' dinner by just saying thank you. I'm very proud of what you've done. It has been an honor and a privilege to work side by side with you to strengthen our democracy. With that I just have two more words to say: Obama out."

Obama then did a mic drop. The crowd of totally objective media members roared with laughter, then rose as one for an extended standing ovation.

> **NPR's review:** "In his roasting session, the president united celebrities, journalists, politicos, and his potential successors by taking them down. Known for his comedic timing and one-liner delivery, Obama didn't disappoint."

> *Washington Post*: "The unusual diversity of the crowd led to some one-of-a-kind encounters. Secretary of State John F. Kerry, Michelle Dockery of 'Downton Abbey,' 'Spotlight' actress Rachel McAdams and MSNBC's Lawrence O'Donnell all gathered at one pre-dinner reception. Vice President Biden sat during the dinner with Kerry and actress Helen Mirren (the vice president typically doesn't attend the dinner; it was Biden's first). At one point Biden greeted singer Gladys Knight by saying, 'I am the Pip!'"

No doubt about it: The press *loved* Barack Obama and Joe Biden. They were just so cool. So hip. And, like, *really* funny!

Most important, they were *Democrats.*

But given the way Team Obama-Biden actually treated the press, why the adulation? Trump may have said mean things about the media and even oftentimes did to their faces, but the preceding occupants of the White House were far, far worse based on something more important:

Their actions.

After all, it wasn't the Trump-Pence administration's Justice Department that obtained a secret warrant and spied on then–Fox News correspondent James Rosen and his parents, it was

Obama-Biden's and their wingman of an attorney general, Eric Holder.

It was Holder who, for no apparent reason, labeled Rosen *an unindicted co-conspirator* and a flight risk in 2013. By doing so, he avoided that pesky need to inform Rosen he was under surveillance.

The reporter, of course, was guilty of absolutely nothing. He was simply pursuing a story and had a source inside U.S. intelligence, who was the target of a Justice Department probe at the time. Rosen was simply the recipient of leaked information from said source.

For that, the Obama-Biden Justice Department affidavit called the reporter "an aider, an abettor, and/or a co-conspirator" who could have ended up being indicted under the Espionage Act.

Glenn Greenwald, one of the most fearless and candid journalists in the game today, was one of the few in the industry who spoke up forcefully at the time.

"Under U.S. law, it is not illegal to publish classified information," Greenwald wrote for the *Guardian* in May 2013. "That fact, along with the First Amendment's guarantee of press freedoms, is what has prevented the U.S. government from ever prosecuting journalists for reporting on what the U.S. government does in secret.

"This newfound theory of the Obama DOJ—that a journalist can be guilty of crimes for 'soliciting' the disclosure of classified information—is a means for circumventing those safeguards and criminalizing the act of investigative journalism itself," he added. "Accusing James Rosen of committing crimes—for basic reporting—may be the most dangerous thing the Obama DOJ has done yet."

A year later, Holder would say that his biggest regret during his time as Obama-Biden's attorney general was the way he handled the Rosen situation. He never apologized to Rosen directly, however, according to the reporter. What a stand-up guy.

Speaking of waging war on the press secretly, it also wasn't Trump-Pence and Attorney General Bill Barr, but rather the Obama-Biden Department of Justice that secretly seized *two months* of phone records from the Associated Press.

"There can be no possible justification for such an overbroad collection of the telephone communications of The Associated Press and its reporters," Gary Pruitt, AP's chief executive officer, wrote at the time. "These records potentially reveal communications with confidential sources across all of the news gathering activities undertaken by The A.P. during a two-month period, provide a road map to A.P.'s news gathering operations, and disclose information about A.P.'s activities and operations that the government has no conceivable right to know."

Strong words. And Pruitt is 100 percent in the right. But oddly, the criticism was short and relatively muted. No extended outrage segments on MSNBC and CNN. No soaring op-eds in the *New York Times* talking about an authoritarian administration acting above the law to chillingly attack the free and fair press.

These stories, to quote the warden in *The Shawshank Redemption*, "up and vanished like a fart in the wind."

Oh, by the way, the Obama administration rejected more Freedom of Information Act requests than any administration in history. This dubious record came eight years after Obama, in his first day in office, pledged to run "the most transparent administration in history."

New York Times reporter James Risen was on target when he called the administration "the most anti-press administration since the Nixon administration."

"Over the past eight years, the administration has prosecuted nine cases involving whistle-blowers and leakers, compared with only three by all previous administrations combined," Risen wrote for the *Times* in a December 30, 2016, column. "It has repeatedly used the Espionage Act, a relic of World War I–era red-baiting, not to prosecute spies but to go after government officials who talked to journalists."

Unfortunately, guys like Risen were on islands when leveling this criticism.

In 2017, after Trump took office, my wife and I attended the White House Correspondents' Association Dinner thanks to the good folks at *The Hill*. We have two young kids and since we both work, elaborate date nights and getting dressed to the nines are opportunities few and far between.

As we entered the Hilton, we were handed First Amendment pins. These were doled out in response to Trump attacking the media for being biased against him. (Such pins were never handed out during the previous administration, which literally spied on reporters and secretly seized phone records. Go figure.)

White House Correspondents' Association president Jeff Mason of Reuters set the tone for the evening, which was strikingly different that the previous eight dinners under the guy who preceded Trump.

"At previous dinners, we have rightly talked about the threats to press freedoms abroad. Tonight we must recognize that there are threats to press freedoms here in the United States. We must remain vigilant. The world is watching."

Trump, wisely, did not attend.

The crowd was also treated to speeches from Bob Woodward and Carl Bernstein, the 1970s journalism icons famous for their Watergate reporting for the *Washington Post*, which took down

President Nixon. If Woodward and Bernstein were a singing duo, they'd be Simon & Garfunkel.

Woodward (Simon) made sure Trump knew that the love fest afforded to Obama and Biden wasn't coming his way.

"The effort today to get this best obtainable version of the truth is largely made in good faith," Woodward said. "Mr. President, the media is not fake news. Let's take that off the table going forward."

Bernstein (Garfunkel) also weighed in: "Almost inevitably, unreasonable government secrecy is the enemy and usually the giveaway about what the real story might be. When lying is combined with secrecy, there is usually a pretty good road map in front of us."

Wow. We're a long way from mic drops and "Obama out."

Joe Biden hasn't been much better with the press as president, either, calling Fox Corporation chairman and News Corp executive chairman Rupert Murdoch, in a conversation with two *New York Times* reporters for their book, "the most dangerous man in the world"

Well, given that Murdoch's *New York Post* was the publication that broke the story about Biden's son Hunter and his laptop, which most of the rest of the Biden-friendly media tried to dismiss, suppress, or outright censor, one can see why Biden would apply such hyperbole.

In the end, however, actions always speak louder than words. Enter Biden's pick to be one of five commissioners of the Federal Communications Commission, Gigi Sohn. Note: The FCC has tremendous power over what you see and hear on television and radio and over speech as a whole.

Here's what Sohn had to say about Fox News in October 2020, a comment that should have automatically disqualified her from even being considered. "For all my concerns about Facebook, I believe that Fox News has had the most negative impact on our democracy.

It's state-sponsored propaganda, with few if any opposing viewpoints. Where's the hearing about that?"

The most negative impact on our democracy? Propaganda? Is this Media Matters or a nominee to sit on the FCC? And what would the goal of this "hearing" be, exactly? An outright ban on cable providers carrying the network?

So it's not just Biden's words, it's his actions. His decision to put such an anti–free speech zealot on such a powerful commission should alarm everyone, but the media (outside of Fox News) didn't condemn Sohn in any way. One could only imagine what the coverage would look like if a Trump appointee said the same thing about CNN:

> IT'S A CHILLING ATTACK ON THE FREE PRESS!
> THIS ISN'T THE SOVIET UNION!

In April 2022, the Biden administration announced that it was launching a "Disinformation Governance Board" to combat what it deems as disinformation. Leading this board would be a woman named Nina Jankowicz, who calls herself "a disinformation fellow" and a Russian disinformation expert.

Here's what Mayorkas's choice to helm Biden's "Ministry of Truth" once said about Hunter Biden's laptop and its damning contents, which many on the left and in the media dubbed Russian disinformation in the weeks before the 2020 election.

"We should view it as a Trump campaign product," Jankowicz said of the story. "Not to mention that the emails don't need to be altered to be part of an influence campaign. Voters deserve that context, not a [fairy] tale about a laptop repair shop," she also tweeted in October 2020.

This was all proven to be completely untrue. Jankowicz was also a big fan of the now-discredited Steele Dossier. Here's what she tweeted about a guest appearance that Christopher Steele made on something called the *Infotagion* podcast: "Listened to this last night. Chris Steele (yes THAT Chris Steele) provides some great historical context about the evolution of disinfo. Worth a listen."

Now she was picked to lead the Ministry of Truth, which wouldn't be politically weaponized or anything. Thankfully, after severe blowback, said Ministry of Truth has been "paused" by Team Biden.

Trump's rhetoric about the press was highly critical and demeaning. Some in the press took it personally and even conflated rhetoric—*Trump's right to free speech*—as somehow endangering their First Amendment.

When looking back on the Trump years, though, it's clear that free speech was more alive and well than ever. It was the most profitable time for journalists and authors in history. Never have more books been written about one president than in the Trump era. And never has more money been made.

Trump. Words.

Obama. Biden. *Actions.*

Which were more harmful to the industry?

That's a rhetorical question.

23

WHO'S AFRAID OF RON DESANTIS?

No active politician scares the Biden administration more than Governor Ron DeSantis of Florida. The choice of coverage by national media from CNN to CBS to the *New York Times* reflects that fear, as no governor in the United States has drawn more national attention. And since DeSantis is a Republican in the mold of Donald Trump, that coverage has been decidedly hostile.

The topic could be DeSantis's handling of COVID. Or his vaccine distribution plans. Or his opening of businesses and schools and beaches while most other states, especially the blue ones, stayed shut down. Or his approving an immigration measure that doesn't allow state entities to do business with companies that transport immigrants who crossed the border illegally into Florida.

Or maybe it's him signing a proclamation saying Emma Weyant was the true winner of a U.S. national college swimming title after she came in second to transgender athlete Lia Thomas, who has a decided size and strength advantage because, you know, she's actually a *he* who suddenly became dominant after jumping from men's competition after being nowhere near the top.

Or perhaps it's his Parental Rights in Education bill (dubbed the "Don't Say Gay" bill by Democrats, echoed by many in the press), a bill that President Biden, who likely didn't read *one sentence of it*, labeled "hateful," while his press secretary at the time, Jen Psaki, called it "heartbreaking."

You can agree or disagree with DeSantis and the Florida legislature on any of these moves or measures or proclamations. What makes the governor popular among his supporters is that he doesn't appear to give a damn about what the Florida press or national political media thinks about how he's leading his state. He has a plan and principles that appear to be unwavering.

Exhibit A is a recent exchange the governor had with WFLA-TV's Evan Donovan after the reporter referenced "what critics call the 'Don't Say Gay' bill."

"Does it say that in the bill?" DeSantis shot back. He knew exactly where this was going and didn't want the reporter to define the bill in the way Democrats wanted it to be framed. "Does it say that in the bill? I'm asking you to tell me what's in the bill because you are pushing false narratives. It doesn't matter what critics say."

"It says 'Classroom instruction on sexual identity and gender orientation,'" Donovan replied while leaving a very key detail out.

"For who? For grades pre-K through three, no five-year-olds, six-year-olds, seven-year-olds," DeSantis retorted without ambiguity. "And the idea that you wouldn't be honest about that and tell people what it actually says, it's why people don't trust people like you because you peddle false narratives. And so we just disabused you of those narratives."

And that's true: The bill applies to kids kindergarten through third grade being taught sexual instruction. To the governor's point, the bill doesn't contain the word *gay* even once. It's also something, being a parent of a kindergartner and second grader myself, that I absolutely support. Because I'm *sane*, as are so many other parents out there regardless of if they voted for Biden or Trump or Bernie or Hillary.

"Understand, if you are out protesting this bill you are by definition putting yourself in favor of injecting sexual instruction to five-, six-, and seven-year-old kids," DeSantis said during another recent press conference. "I think most people think that's wrong. I think parents especially think that's wrong."

The national press, on cue, is largely against the bill, and each outlet made sure to call it the "Don't Say Gay" bill in every headline to drive the farcical narrative home.

> NBC News: "Florida Governor Ron DeSantis Signals Support for 'Don't Say Gay' Bill: The bill, which would bar the 'discussion of sexual orientation or gender identity' in primary schools, passed the Florida Senate Education Committee on Tuesday."

The headline itself is misleading, because *that's not what the bill is called*, it's what critics from the left and activists call it. Throughout this entire NBC News story that isn't labeled an opinion piece (but should be), never once does it mention DeSantis's main point: that the bill bars sexual instruction to five-, six-, and seven-year-old kids. Why omit that crucial element to the legislation? Unless, of course, fiction is being presented under the guise of news.

Despite all of the negative press around it, Floridians by a wide margin support the bill as it pertains to "banning the teaching of sexual orientation and gender identity from kindergarten through third grade." Quinnipiac polling in March 2022 found that just 35 percent oppose the bill.

Biden and Psaki and the press failed. Because, thankfully, more and more, people can spot bullshit when they hear it.

DeSantis, dubbed the "Angel of Death" by *Vanity Fair* and "DeathSantis" by several liberal hosts and publications, is leading his Democratic challengers in this year's Florida's governor's race by considerable margins and is set to not only win a second term, but have a clear path to the Republican presidential nomination if Trump doesn't run.

Overall, DeSantis, an Iraq War veteran and Harvard Law graduate, sat comfortably above 50 percent approval in Florida while President Biden sits comfortably below 40 percent in most polls. Double whoops.

When 2024 rolls around, Donald Trump will be seventy-eight years old. DeSantis will be forty-five. A June 2022 straw poll at the Western Conservative Summit showed DeSantis topping Trump with 71 percent of the vote to Trump's 67 percent (those voting could choose multiple candidates). DeSantis certainly has momentum going into 2023. The question is: Could he beat Trump in a head-to-head matchup? His chances may be better than you think, and Trump knows it.

DeSantis's positions and perspectives and demeanor may be hated by Democrats and the press, and they certainly make that well known. This chapter serves as a reminder and possible preview for 2024: The vitriolic posture of many in the press didn't just come along with Donald Trump. It dates back to Reagan and the Bushes, albeit to a much lesser extent. John McCain was supposed to be the guy who got a fair shake in 2008 when he ran for president after building such a solid rapport and reputation with the media. That all went away when Barack Obama came along, and suddenly McCain's portrayal went from dignified war hero to racist curmudgeon.

In August 2008, the *New York Times* editorial page described an official McCain ad as "racially tinged" because it included a photo of Obama juxtaposed with Paris Hilton and Britney Spears.

The message the McCain campaign was sending: Obama, with little experience but on the receiving end of media adulation, was much more sizzle than steak. Fair criticism.

No matter: Ezra Klein wrote in the *American Prospect* at the time that the McCain campaign was "running crypto-racist ads." Bill Press called it "deliberately and deceptively racist." Blogger Josh Marshall, a winner of the prestigious George Polk Award, argued that "the McCain campaign is now pushing the caricature of Obama as an uppity young black man whose presumptuousness is displayed not only in taking on airs above his station but also in a taste for young white women." CNN's Don Lemon accused the campaign of "creating a political environment that is inciting hate and hate speech."

After McCain lost, all of this criticism of him magically went away.

Because mission accomplished in helping Barack Obama and Joe Biden take office.

The 2012 election also featured a nice guy capturing the Republican nomination in Mitt Romney. In his treatment to this day, from CNN's Candy Crowley inappropriately jumping in to defend Obama during a crucial debate only to admit (after the damage was done) that the Republican's argument was actually correct, to Romney being accused of pooch abuse while also being cast as a closet racist dressed as Gordon Gekko, the same playbook is applied:

The nominee is a racist. He's also a hateful person. This approach will be used again and again, whether it's Trump, DeSantis, Tim Scott, Nikki Haley, Kristi Noem, or the second coming of Abraham Lincoln at the top of the ticket in 2024 against Biden or (enter nominee here).

CONCLUSION

In the final days of the 1980 presidential campaign, Republican Ronald Reagan asked Americans a devastating question about incumbent Jimmy Carter.

"Are you better off than you were four years ago?"

The answer, of course, was no. Inflation and gas prices were soaring. The economy was experiencing negative growth. Fifty-two Americans were still being held hostage by Iran after a rescue mission ordered by Carter failed miserably, resulting in eight U.S. servicemen dead and no hostages retrieved. America wasn't feeling very good about itself heading into Election Day 1980, yet the race was still close when the first and only debate between the former actor and former peanut farmer took place on October 28.

"Next Tuesday all of you will go to the polls, [you] will stand there in the polling place and make a decision," Reagan said that night with 81 million people watching at home.

> I think when you make that decision, it might be well if you would ask yourself, are you better off than you were four years ago? Is it easier for you to go and buy things in the stores than it was four years ago? Is there more or less unemployment in the country than there was four years ago? Is America as respected throughout the world as it was? Do you feel that our security is as safe, that we're as strong as we were four years ago? And if you answer all of those questions yes, why then, I think your choice is very obvious as to whom you will vote for.

> If you don't agree, if you don't think that this course that we've been on for the last four years is what you would like to see us follow for the next four, then I could suggest another choice that you have.

He added with pitch-perfect delivery, "This country doesn't have to be in the shape that it is in."

These are the same basic questions we face before the next midterms and the next presidential election:

Are you better off than you were before Joe Biden took office?

No.

Can you buy things in stores easier under this president?

No.

Is unemployment lower than it was entering 2021?

Yes. *But* . . . all that's happening is jobs being *added back* to what were lost during the pandemic, when businesses were forcibly shuttered or destroyed. This isn't job creation, it's job *restoration.* And more Americans are quitting their jobs than ever before.

Is America as respected throughout the world as it was before Joe Biden?

Hell no.

Do you feel that our security is as safe, that we're as strong as we were under the previous administration?

Hell no again.

For whoever the nominee is for the Republican Party in 2024 is, the campaign should be laser-focused on two things:

1. Making the election a referendum on Biden's performance as president by asking two questions: Are you better off, or worse off, under this president? And are you confident as Biden goes further into his eighties that he'll improve in his ability to run the country? These are two very big questions with obvious answers to anyone sober and paying attention.
2. Offering solutions anchored in common sense. As Hillary learned, you can't just say, "That guy sucks," and expect to win. A new Contract with America must be offered under the KISS (Keep It Simple, Stupid) principle.

The playbook is already available in the form of the masterful campaign that rookie politician Glenn Youngkin ran in the Virginia governor race in 2021. Because despite trailing by double digits heading into the final month of the campaign against Clinton royalty Terry McAuliffe, Youngkin tapped into the issues that residents of the commonwealth who are not blindly married to one political party cared about most: Education. Economy. Removing COVID restrictions. Election integrity.

Youngkin won because he brought suburban momma bears back over to the GOP; they were energized on education issues particularly as they pertained to Critical Race Theory being taught in schools, and school boards (and McAuliffe) telling parents to sit down and shut up when it came to having a say in *their child's education.*

"The top nine most terrifying words in the English language are: I'm from the government, and I'm here to help," Reagan once famously said.

Joe Biden and his party have embraced the exact opposite in believing that more government is the solution, not the problem. Remember, as much as Biden seems to be fumbling, all his fumbles go leftward because that's who he is. As catastrophic as his first year has been, it's not an *accidental* catastrophe. It's Joe Biden's fault. It's a catastrophe stemming directly from the sort of choices and signals the Biden administration sends out. This isn't Bill Clinton in 1996, who wisely pivoted to the center in working with the new Republican majority, led by Newt Gingrich in the House. Could you honestly ever imagine Joe Biden, flanked by AOC and Nancy Pelosi, declaring that "the era of big government is over," as Clinton did in advocating welfare reform?

Never. CNN and MSNBC talking heads and the woke social media mob would never allow for such a pivot. Moving forward, Biden will continue to embrace the left. Voters, including Blue Dog Democrats, independents, and (some) Republicans offended by Trump, will all come back to the GOP in the midterms.

The same left will turn on Biden in 2023, all while not understanding that *their* policies, *their* worldview, was why they lost power before. It's pretty easy to predict that Biden will announce he won't be seeking a second term, creating an open primary with few good options for the party.

We've heard it so many times before, but this time this really is the deal for 2024: It will mark the most important election of our lifetimes.

Because America has just witnessed what the Democratic Party ran on in 2020 and attempted to do in 2021 and 2022, when it thought it could ram through whatever it wanted in the House and Senate, until Manchin and Sinema said otherwise:

- The biggest expansion of government in history
- Spending trillions upon trillions while inflation and the deficit were already at a forty-year high
- Turning our educational system over to the 1619 Project and CRT and extremists on sexual orientation teaching. Kindergartners being programmed to see the world through race while being pushed to question if they're a boy or a girl.

As a dad of two young kids, I see the world through a different prism than I did in my pre-kids world. I'm genuinely scared for the country.

We spend what we don't have.

We have a border czar and DHS secretary who are allowing fentanyl to kill Americans at an alarming rate, and they simply do not care.

We don't have the right people in place to fix inflation or the supply chain.

Those in charge seem only interested in holding on to power.

We allow our adversaries, China, Russia, Iran, the Taliban, to dictate our actions and those of the world. We allow our own border to be overrun while allowing deadly fentanyl in. We *don't* allow our kids to be kids anymore. Instead, it's some kind of bizarre social experience that takes the focus off the basic blocking and tackling they need to be prepared for adulthood: reading, writing, math, science.

The country is decidedly on the wrong track. The year 2024 is a chance to turn all of that around, with the midterms being Part 1 of a two-part series.

Joe Biden will go down as the worst president of our lifetime. Yet many in our media will continue to prop him up for as long as possible. Too many journalists have forsaken objectivity. When he leaves the stage, they'll seek an even more extreme candidate, assuming that his failures were due to his "normalcy" rather than his extreme policies. They'll embrace someone even farther left. Social media will do their part as well. The fix will be in.

But I have faith that a solid majority of the American people will decipher fact from opinion and partisan grandstanding. There are so many options for those on the right and left and middle. And more speech is better than less speech.

The late Tim Russert has been praised in this book for good reason. Before his untimely death in 2008, *Meet the Press* truly was a place where you felt like the moderator was interested in challenging those in power. It didn't matter if there was a "D" or an "R" next to their names.

In 2004, following the presidential election, Russert gave a prophetic interview at the JFK Library. He saw where the industry was going as the internet became mainstream. And he embraced it.

"The information spectrum has exploded," Russert said. "There are the three major networks, there's twenty-four-hour cable, there's the talk radio, the internet, the bloggers, everybody.

"But I think we have to be very careful that we don't become afraid of that. It's nothing to be afraid of. It is what it is, it's real, and it's part of our life, and the American people understand that. They know when they see someone on *Meet the Press* it's different than watching someone on *O'Reilly* or on *Larry King*. They really have a very good radar detector, and they know that Rush Limbaugh has a point of view and Al Franken has a point of view."

Wow. Faith in the consumer to cut through the noise. Novel concept.

Oh, and here's the best part. There's this great invention called the remote control. It allows you to shut off something you don't like, or simply change the channel.

So the next time you feel like you're being lectured to. Or lied to. Or see an anchor and journalist clearly rooting for one side and against another, look for the most important remote button of them all:

Off.

ACKNOWLEDGMENTS

Truth be told, I never thought I'd be writing acknowledgments for any book. When it comes to writing, I've always been more of a sprinter then a long-distance runner. And to be totally candid, I didn't think I had the time or the discipline to finish one of these things, particularly with my kids keeping my wife and me so active that Jean and I usually pass out by 9:15 every night.

But you see the crazy direction that this country is going in, and suddenly that time becomes available and that motivation to discipline oneself to go to a library for three hours a day becomes natural.

I'd like to first extend my appreciation to executive Eric Nelson at HarperCollins. His advice before I started this journey was so spot-on at a time the task seemed so daunting. "Write the interesting parts first," he recommended. "If you're eager to lay into a topic, but it's chapter six, still, write it first. The parts that are fun to write will be the best parts of the book, and the parts that are a slog to write probably won't be enjoyable to read in the end (so consider cutting or reducing the latter)." Genius stuff.

The editing and manuscript responsibilities were handled seamlessly and professionally by Hannah Long and James Neidhardt, and these two didn't miss a thing while keeping everything on deadline. If I write another book down the road, I'll make my first demand working with this team again.

Next, I want to thank my young children, Cameron and Liam. They were only in second grade and kindergarten at the time I was writing this. And when my son wanted to go throw a ball around or my daughter wanted to go for a bike ride, sometimes I had to tell them I just didn't have any time. It hurt to do that, but now that I have that time back it's appreciated even more.

My dad deserves kudos as well for providing the usual candid feedback on the book as I sent bits and pieces to him along the way. I know I'm always going to get honest assessment, good or not so good.

Finally, I'd like to thank my wife for helping me find the hours and doing more than 110 percent she's already doing in raising our kids right. Jean is an ER doc. In the medical profession, it's arguably the toughest job of them all. The hours are long. Oftentimes she has to work weekends or early in the morning or late into the evening. But she allowed me to follow my passion, and words don't express my appreciation enough.

INDEX